T0100035

NANOTECHNOLOGY SCIENCE AND TECHNOLOGY

ELECTROSPINNING PROCESS AND NANOFIBER RESEARCH

NANOTECHNOLOGY SCIENCE AND TECHNOLOGY

Additional books in this series can be found on Nova's website under the Series tab.

Additional E-books in this series can be found on Nova's website under the E-book tab.

NANOTECHNOLOGY SCIENCE AND TECHNOLOGY

ELECTROSPINNING PROCESS AND NANOFIBER RESEARCH

A.K. HAGHI
AND
G.E. ZAIKOV
EDITORS

Nova Science Publishers, Inc.
New York

Library of Congress Cataloging-in-Publication Data

Electrospinning process and nanofiber research / editors, A.K. Haghi, G.E. Zaikov.
p. cm. -- (Nanotechnology science and technology)
Includes bibliographical references and index.
ISBN 978-1-61209-330-7 (hardcover : alk. paper)
1. Tissue engineering. 2. Nanofibers. 3. Electrospinning. I. Haghi, A. K. II. Zaikov, G. E. (Gennadii Efremovich), 1935-
R857.T55E43 2011
610.28--dc22

2010048371

Published by Nova Science Publishers, Inc. † New York

Contents

PREFACE

Electrospinning is the cheapest and the most straightforward way to produce nanomaterials. Electrospun nanofibres are very important for the scientific and economic revival of developing countries. Electrospinning was developed from electrostatic spraying and now represents an attractive approach for polymer biomaterials processing, with the opportunity for control over morphology, porosity and composition using simple equipment. Because electrospinning is one of the few techniques to prepare long fibres of nano- to micrometre diameter, great progress has been made in recent years.

It is now possible to produce a low-cost, high-value, high-strength fibre from a biodegradable and renewable waste product for easing environmental concerns. For example, electrospun nanofibres can be used in wound dressings, filtration applications, bone tissue engineering, catalyst supports, non-woven fabrics, reinforced fibres, support for enzymes, drug delivery systems, fuel cells, conducting polymers and composites, photonics, medicine, pharmacy, fibre mats serving as reinforcing component in composite systems, and fibre templates for the preparation of functional nanotubes.

This update covers all aspects of electrospinning as used to produce Nanofibres.

In: Electrospinning Process and Nanofiber... ISBN 978-1-61209-330-7
Editors: A.K. Haghi and G.E. Zaikov

Chapter 1

PREDICTION OF NANOFIBER DIAMETER FOR IMPROVEMENTS IN INCORPORATION OF MULTILAYER ELECTROSPUN NANOFIBERS

A. K. Haghi[*]
University of Guilan, Rasht, Iran

ABSTRACT

Electrospun nanofiber web has many potential applications due to its large specific area, very small pore size and high porosity. Despite such potentials, the mechanical properties of nanofiber web are very poor for use in textile application.

To remedy this defect, laminating process could accomplish in order to protect nanofiber web versus mechanical stresses. In this paper, direct tracking method as an image analysis based technique for measuring electrospun nanofiber diameter has been presented. The usefulness of the method for electrospun nanofiber diameter measurement discussed. Such automated measurement of nanofiber diameter can be used to obtain better laminated webs.

Keywords: Laminated nonwovens, Electrospinning, Nanofibers, Fiber diameter, Image analysis.

---

[*] Haghi@Guilan.ac.ir

1. INTRODUCTION

In Electrospinning process, a high electric field is generated between a polymer solution held by its surface tension at the end of a syringe (or a capillary tube) and a collection target.

In the fabric lamination, producing an adhesive bond which guarantees no delaminating or failure in use requires lamination skills and information about adhesive types. It is relatively simple to create a strong bond; the challenge is to preserve the original properties of the fabric and to produce a flexible laminate with the required appearance, handle and durability. In other words, the application of adhesive should have minimum affect on the fabric flexibility and aesthetics during the lamination process 63, therefore, adhesive must be applying in a controlled manner. In order to achieve to this purpose, it is generally necessary that the least amount of a highly effective adhesive applied and it penetrate to a certain extent of the fabric and cover the widest possible surface area. Too much adhesive and excessive penetration likely to lead to fabric stiffening and it could result in thermal discomfort in the cloth; since the adhesive itself could form an impermeable barrier to perspiration.

The adhesives could be as solvent/water-based adhesive or as hot-melt adhesive. In first group, the adhesives are as solutions in solvent or water, and solidify by evaporating of the carrying liquid.

In this group, solvent-based adhesives could 'wet' the surfaces to be joined better than water-based adhesives, and also could solidify faster. But unfortunately, they are environmentally unfriendly, usually flammable and more expensive than those. Of course it's not means that the water-based adhesives are always preferred to laminating, since in practice, drying off water in terms of energy and time is expensive too. Beside, water-based adhesives are not resisting to water or moisture because of their hydrophilic nature. But in hot-melt adhesive group, the adhesives are as solids and melt under the action of heat. These types of adhesives are environmentally friendly, inexpensive, require less heat and energy, and so is now more preferred. They can be of several different chemical types, such as polyolefin (polyethylene, polypropylene), polyurethane, polyester, polyamide or blends of different polymers or copolymers in order to reach for a wide range of properties (including melting points, durability to washing and dry cleaning and heat resistance). Hot-melt lamination can be either continuous (hot calendars) or static (flat iron or Hoffman press) and is accomplished by two separate processes: first a means of applying the actual adhesive; and second bringing the two substrates together to form the actual bond under the action of heat and pressure. In this process, the heating accomplish at temperatures above the softening or melting point of adhesive.

In addition, hot melt adhesives are available in several forms; as a web, as a continuous film, or in powder or granular form. The adhesive powders are available in most chemical types and also in particle sizes ranging from very small up to about 500 micrometers or so in diameter. Adhesives in film or web form are more expensive than the corresponding adhesive powders. The webs are discontinuous and produce laminates which are flexible, porous and breathable, whereas, Continuous film adhesives cause stiffening and produce laminates which are not porous and permeable to both air and water vapor. This behavior attributed to impervious nature of adhesive film and its shrinkage under the action of heat.

Figure 1 represents the optical microscope image of multilayer nanofiber web. Accurate and automated measurement of nanofiber diameter of laminated webs is useful and crucial and therefore has been taken into consideration in this contribution. The objective of the current research would then be to develop an image analysis based method to serve as a simple, automated and efficient alternative for electrospun nanofiber diameter measurement with particular application in laminated nanofiber web.

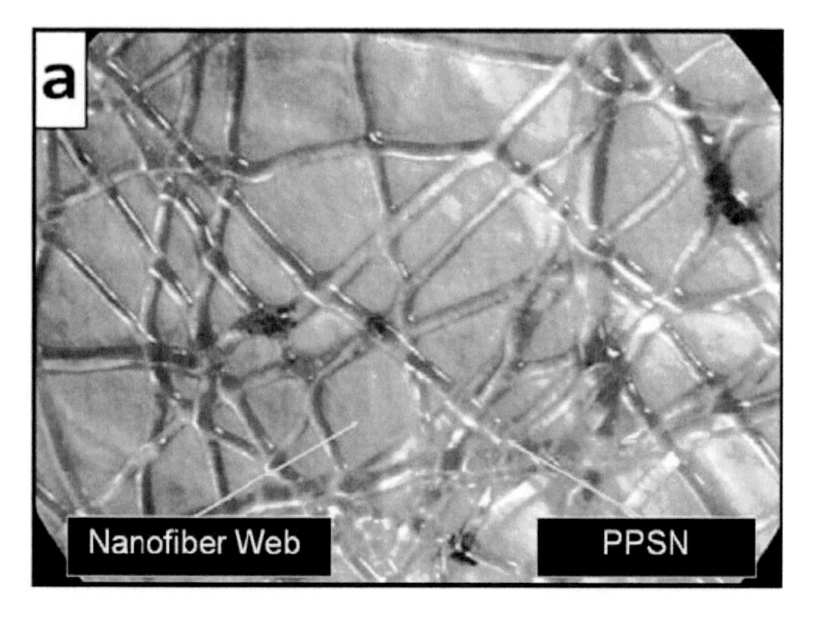

Figure 1. The optical microscope images of multilayer nanofiber web (PPSN) : polypropylene spun-bond nonwoven.

2. Methodology

The algorithm for determining fiber diameter uses a binary input image and creates its skeleton and distance transformed image (distance map). The skeleton acts as a guide for tracking the distance transformed image and fiber diameters are measured from the intensities of the distance map at all points along the skeleton . Figure 1. shows a simple simulated image, which consists of five fibers with diameters of 10, 13, 16, 19 and 21 pixels, together with its skeleton and distance map including the histogram of fiber diameter obtained by this method. In this paper, we developed *direct tracking* method for measuring electrospun nanofiber diameter. This method which also uses a binary image as the input, determines fiber diameter based on information acquired from two scans; first a horizontal and then a vertical scan. In the horizontal scan, the algorithm searches for the first white pixel (representative of fibers) adjacent to a black (representative of background).

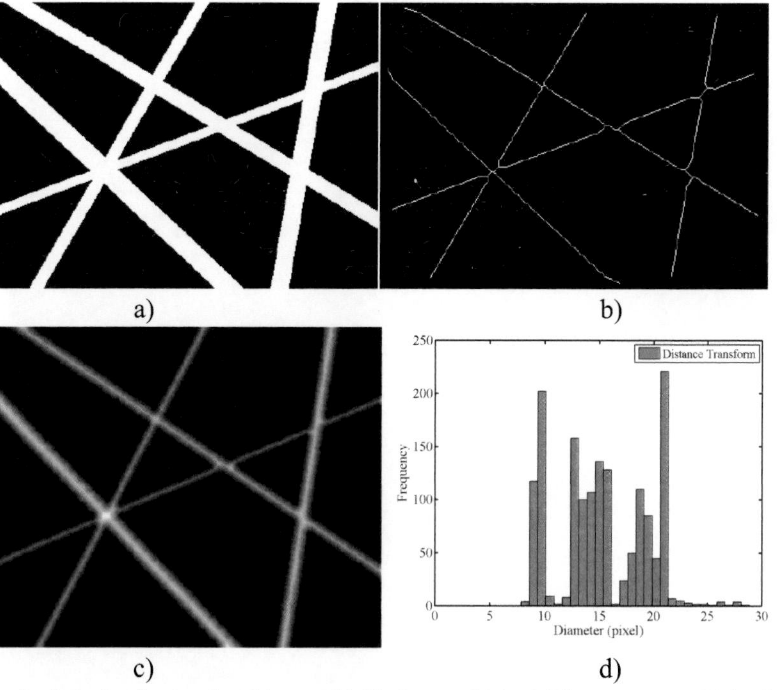

Figure 2. a) A simple simulated image, b) Skeleton of (a), c) Distance map of (a) after pruning, d) Histogram of fiber diameter distribution obtained by distance transform method.

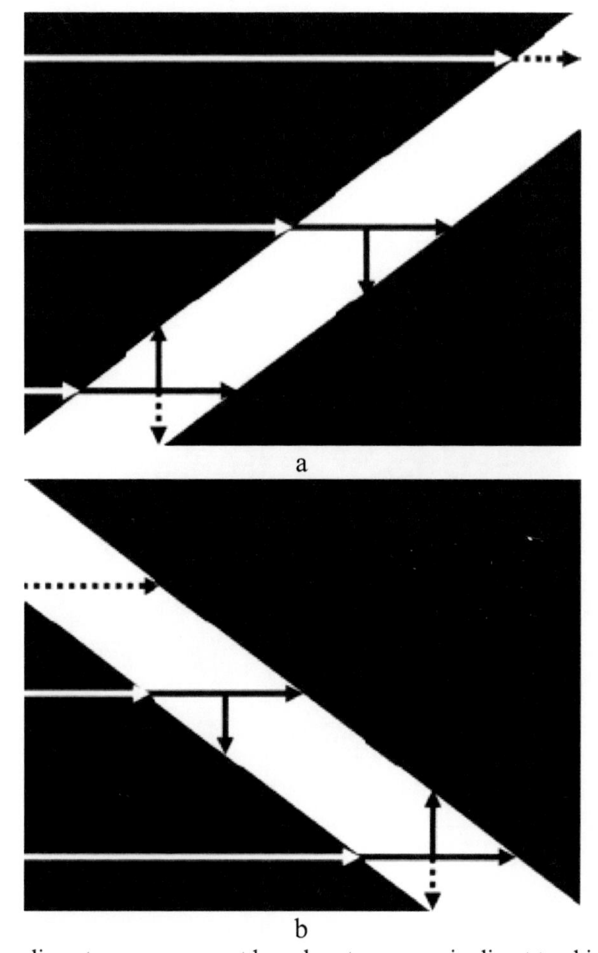

Figure 3. Fiber diameter measurement based on two scans in direct tracking method.

Pixels are then counted until reaching the first black. Afterwards, the second scan is started from the mid point of horizontal scan and pixels are counted until the first vertical black pixel is encountered. Direction will change if the black pixel isn't found (Figure 2). Having the number of horizontal and vertical scans, the number of pixels in perpendicular direction which is the fiber diameter in terms of pixels can be measured through a simple geometrical relationship.

In electrospun webs, nanofibers cross each other at intersection points and this brings about the possibility for some untrue measurements of fiber diameter in these regions. To circumvent this problem, a process called *fiber identification* is employed. First, black regions are labeled and a couple of regions between which a fiber exists, are selected. Figure 3. depicts the labeled simulated image and the histogram of fiber diameter obtained by direct tracking method.

Now, reliable evaluation of the accuracy of the developed methods requires samples with known characteristics. Since it is neither possible to obtain real electrospun webs with specific characteristics through the experiment nor there is a method which measures fiber diameters precisely with which to compare the results, the method will not be well evaluated using just real webs.

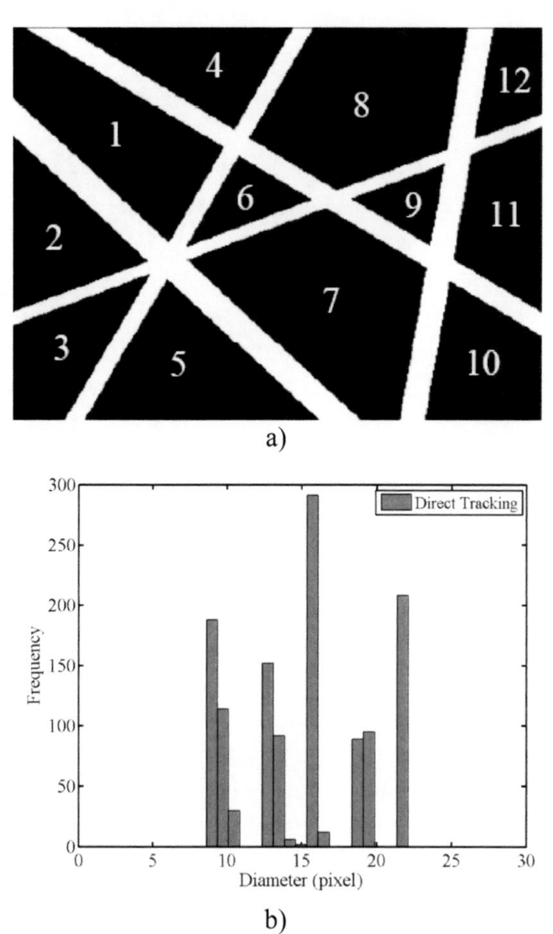

Figure 4. a) The labeled simulated image, b) Histogram of fiber diameter distribution obtained by direct tracking method.

To that end, a simulation algorithm has been employed for generating samples with known characteristics. In this case, it is assumed that the lines are infinitely long so that in the image plane, they intersect the boundaries. Under this scheme, which is shown in Figure 4, a line with a specified thickness is defined by the perpendicular distance d away from a fixed reference point O located in the center of the image and the angular position of the perpendicular α. Distance d is limited to the diagonal of the image . Several variables are allowed to be controlled during simulation; line thickness, line density, angular density and distance from the reference point. These variables can be sampled from given distributions or held constant. Distance transform and direct tracking algorithms for measuring fiber diameter both require binary image as their input. Hence, the micrographs of electrospun webs first have to be converted to black and white. This is carried out by *thresholding* process (known also as *segmentation*) which produces binary image from a grayscale (intensity) image . This is a critical step because the segmentation affects the result significantly. Prior to the segmentation, an *intensity adjustment* operation and a two dimensional *median* filter are often applied in order to enhance the contrast of the image and remove noise. In the simplest thresholding technique, called *global thresholding*, the image is segmented using a single constant threshold. One simple way to choose a threshold is by trial and error. Each pixel is then labeled as object or background depending on whether its gray level is greater or less than the value of threshold respectively. The main problem of global thresholding is its possible failure in the presence of non-uniform illumination or local gray level unevenness. An alternative to this problem is to use *local thresholding* instead. In this approach, the original image is divided to subimages and different thresholds are used for segmentation. As it is shown in Figure 5, global thresholding resulted in some broken fiber segments. This problem was solved using local thresholding.

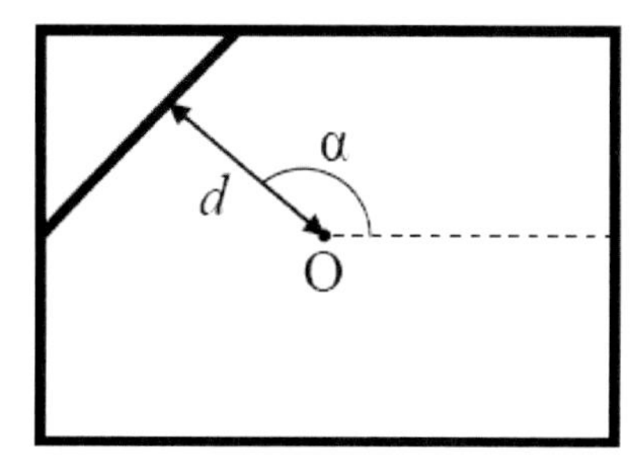

Figure 5. μ-randomness procedure.

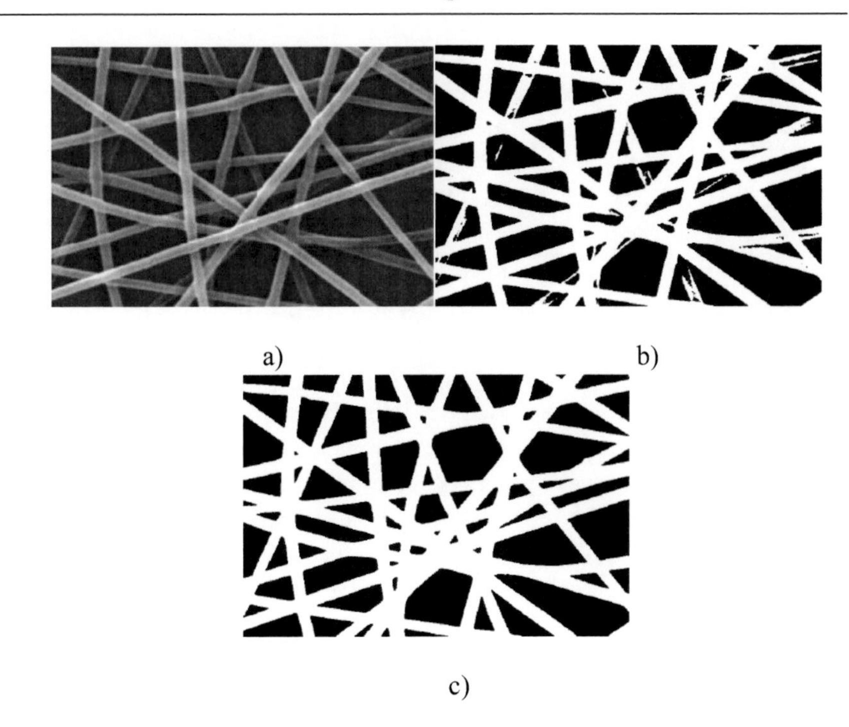

Figure 6. a) A typical electrospun web, b) Global thresholding, c) Local thresholding.

3. Experimental

Electrospun nanofiber webs used as real webs in image analysis were prepared by electrospinning aqueous solutions of PVA with average molecular weight of 72000 *g/mol* (MERCK) at different processing parameters. The micrographs of the webs were obtained using a Philips (XL-30) environmental Scanning Electron Microscope (SEM) under magnification of 10000X after gold sputter coating.

4. Results and Discussion

Three simulated images generated by μ-randomness procedure were used as samples with known characteristics to demonstrate the validity of the techniques. They were each produced by 30 randomly oriented lines with varied diameters sampled from normal distributions with mean of 15 pixels and standard deviation of 2, 4 and 8 pixels respectively. Table 1 summarizes the structural features of these simulated images which are shown in Figure 6.

Table 1. Structural characteristics of the simulated images generated using μ-randomness procedure

No.	Angular range	Line density	Line thickness	
			M	Std
1	0-360	30	15	2
2	0-360	30	15	4
3	0-360	30	15	8

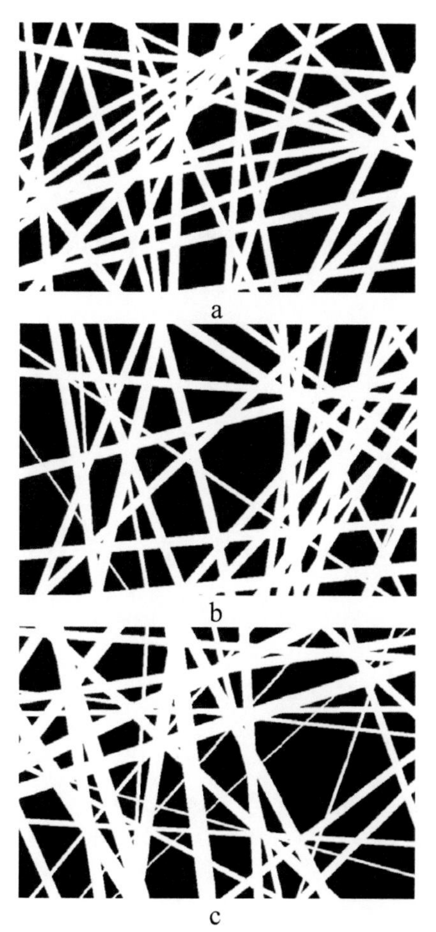

Figure 6. Simulated images generated using μ-randomness procedure.

Table 2. Mean and standard deviation of fiber diameters for the simulated images

		No. 1	No. 2	No. 3
Simulation	M	15.247	15.350	15.367
	Std	1.998	4.466	8.129
Distance transform	M	16.517	16.593	17.865
	Std	5.350	6.165	9.553
Direct tracking	M	16.075	15.803	16.770
	Std	2.606	5.007	9.319

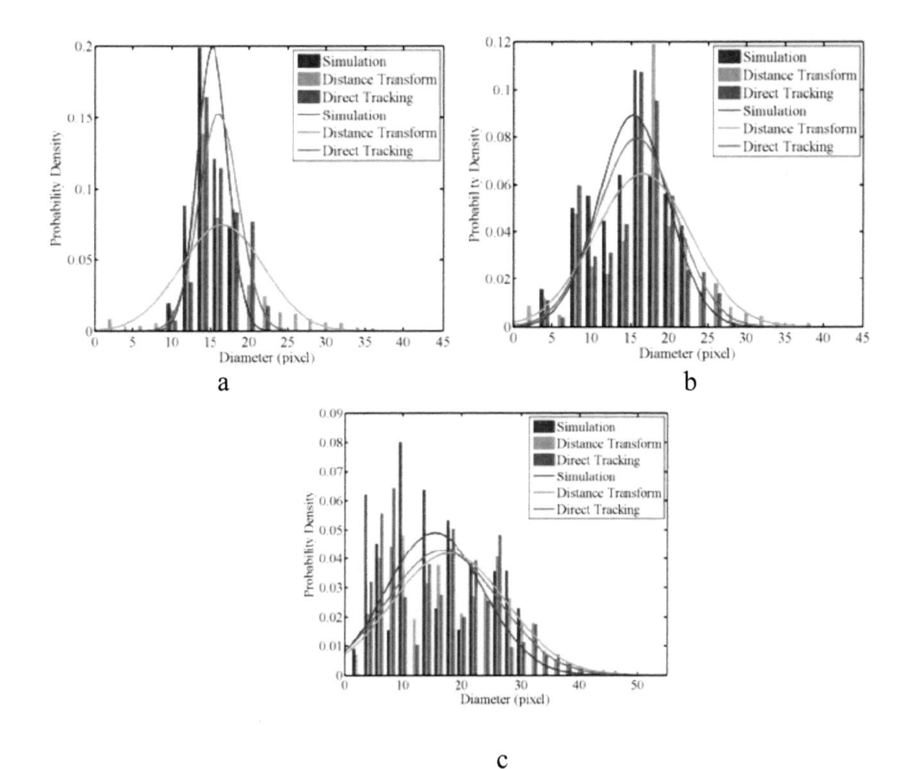

Figure 7. Histograms of fiber diameter distribution for the simulated images.

Mean and standard deviation of fiber diameters for the simulated images obtained by direct tracking as well as distance transform are listed in Table 2. Figure 7. shows histograms of fiber diameter distribution for the simulated images obtained by the two methods. In order to make a true comparison, the original distribution of fiber diameter in each simulated image is also included.

The line over each histogram is related to the fitted normal distribution to the corresponding fiber diameters.

Table 2. and Figure 7 clearly demonstrate that for all simulated webs, direct tracking method resulted in mean and standard deviation of fiber diameters which are closer to those of the corresponding simulated image (the true ones). Distance transform method is far away from making reliable and accurate measurements. This may be due to remaining some branches in the skeleton even after pruning. The thicker the line, the higher possibility of branching during skeletonization (or thinning). Although these branches are small, their orientation is typically normal to the fiber axis; thus causing widening the distribution obtained by distance transform method.

Conclusion

Fiber diameter is one of the most important structural characteristics in electrospun nanofiber webs. Electrospun nanofiber diameter is often measured by manual method – labor intensive, time consuming, operator-based technique which only utilizes a low number of measurements, thereby inefficient for automated systems e.g. online quality control. In this study, an automated technique called "Direct Tracking" for measuring electrospun nanofiber diameter has been developed that is fast and has the capacity for automation, enabling improved quality control of large scale electrospinning operations.

References

M. Ziabari, V. Mottaghitalab, A.K.Haghi, Application of direct tracking method for measuring electrospun nanofiber diameter. *Braz. J. Chem. Eng.* , v. 26, n. 1, pp. 53-62, 2009.

M. Ziabari, V. Mottaghitalab, S. T. McGovern, A. K. Haghi, Measuring Electrospun Nanofibre Diameter: A Novel Approach, *Chin.Phys.Lett.* , Vol. 25, No. 8 , pp. 3071-3074, 2008.

M. Ziabari, V. Mottaghitalab, S. T. McGovern, A. K. Haghi, A new image analysis based method for measuring electrospun nanofiber diameter, *Nanoscale Research Letter* , Vol. 2, pp. 297-600, 2007.

M. Ziabari, V. Mottaghitalab, A. K. Haghi, Simulated image of electrospun nonwoven web of PVA and corresponding nanofiber diameter distribution, *Korean Journal of Chemical engng* , Vol.25, No. 4, pp. 919-922 ,2008.

M. Ziabari, V. Mottaghitalab, A. K. Haghi, Evaluation of electrospun nanofiber pore structure parameters, *Korean Journal of Chemical engng* , Vol.25, No. 4, pp. 923-932 ,2008.

M. Ziabari, V. Mottaghitalab, A. K. Haghi, Distance transform algoritm for measuring nanofiber diameter, *Korean Journal of Chemical engng*, Vol25, No. 4, pp. 905-918 ,2008.

A.K.Haghi and M. Akbari, Trends in electrospinning of natural nanofibers, *Physica Status Solidi* , Vol.204, No. 6 pp. 1830–1834 , 2007.

M. Ziabari, V. Mottaghitalab, A. K. Haghi, A new approach for optimization of electrospun nanofiber formation process, *Korean Journal of Chemical engng,* DOI: 10.2478/s11814-009-0309-1.

In: Electrospinning Process and Nanofiber... ISBN 978-1-61209-330-7
Editors: A.K. Haghi and G.E. Zaikov

Chapter 2

ELECTROSPINNING OF HIGH CONCENTRATION GELATIN SOLUTIONS

Tudorel Balau Mindru[1], Iulia Balau Mindru[1], Theodor Malutan[2] and Vasile Tura[3]

[1]Faculty of Textiles and Leather Engineering, Gh. Asachi Technical University, Str. Dimitrie Mangeron nr. 53, Corp TEX1, 700050, Iasi, Romania
[2]Faculty of Industrial Chemistry, Gh. Asachi Technical University, Str. Dimitrie Mangeron nr. 71, 700050, Iasi, Romania
[3]Faculty of Physics, Al. I. Cuza University, Blvd. Carol I, nr. 11A, 700506, Iasi, Romania

ABSTRACT

Non-woven gelatin nanofiber membranes were prepared by electrospinning of high concentration gelatin solutions. Electrospinnable solutions of 27-30% (w/v) gelatin concentration were prepared using various solvents. The solvent mixture consisting of formic acid, acetic acid and dimethylformamide in 4:0.5:0.5 volume ratio gave the best results, the gelatin solution remaining stable for more than 48 hours, as proved by viscosity measurements. The chemical and physical structure of the gelatin nanofiber membranes were investigated by scanning electron microscopy, infrared spectroscopy, differential scanning calorimetry and X-ray diffraction. The observed chemical and physical properties were compared to those of gelatin films prepared from the same solutions. The films showed structural differences depending on the solvent mixture used, while the characteristics of the electrospun membranes were almost similar.

Keywords: Electrospinning; Gelatin; Nanofibers.

1. INTRODUCTION

Gelatin is a natural biopolymer with a wide range of applications in medical, pharmaceutical and food industries. In the last years, gelatin based scaffolds prepared by electrospinning were intensely investigated because their three-dimensional (3D) structure that mimics very well the extracellular matrix makes them very attractive for tissue engineering applications [1-5]. The first important step in electrospinning a natural polymer is the preparation of an electrospinnable solution using a proper solvent. Gelatin is a natural polymer with strong polarity. It has molecular chains connected through strong hydrogen bonds, constituting a 3D macromolecular network (double or triple helix) with reduced mobility [6]. In order to dissolve gelatin, high-polarity solvents are required to break the links between chains and change its structure from helix to random-coil. Gelatin can be made to undergo reversible helix-coil transformations by choosing appropriate solvent systems. Addition of urea or thiocyanates to aqueous gelatin solutions hinders gelatin renaturation, and treating films of helical gelatin with solutions of urea or thiocyanates results in a transition to the coil structure [7]. Adding crosslinking agents to a gelatin solution may promote or hinder renaturation of gelatin, depending on the relative rates of the helix formation and crosslinking processes [8]. Water is a good solvent of gelatin because it breaks very easy the interchain links and produces stable solutions at 50°C, without gelatin degradation. However, an aqueous gelatin solution changes into a gel in the syringe needle at room temperature and the electrospinning process becomes impossible. Moreover, water has a slow evaporation rate, making impossible the transformation of the solution filament into a dry nanofiber during the travel between needle and collector. Because of these limitations in using water as a solvent, electrospinning of gelatin requires the use of fast-evaporating organic solvents. Formamide, dimethyl sulphoxide and 2-chloroethanol were used because they prevent helix formation [8]. Fluorinated alcohol solvents such as trifluoroethanol and hexafluoro isopropanol were found also to be good solvents for polypeptides [9]. The formic acid was found to produce gelatin solutions suitable for electrospinning experiments. Solutions of 7-12% (w/v) gelatin dissolved in formic acid were successfully electrospun into nanofibers with diameters in the range from 70 nm to 170 nm [10]. But, in time, the formic acid determines gelatin degradation and an important decrease of the solution viscosity, which makes the electrospinning process impossible due to beads formation [10]. The second important problem in gelatin electrospinning is the gelatin concentration of the electrospinnable solution. It was reported that gelatin solutions with gelatin concentrations higher than 12% could not be electrospun because a hardened gelatin phase developed on the edge of the needle tip, disturbing the fluid filament flow and the quality of the resulted nanofibers [3].

However, electrospinning of high concentration gelatin solutions is desirable, because the initial gelatin concentration was shown to be the most important parameter in controlling the cross-linking density and cytotoxin formation in gelatin biomedical applications [11, 12]. In the present work, we investigated the possibility to increase the gelatin concentration of electrospinnable solutions by using mixtures of formic acid (FA), acetic acid (AA) and dimethylformamide (DMF) in suitable ratios. The optimized solvent mixture was found to extend the solution stability interval, also.

2. Experimental

2.1. Gelatin Preparation

A gelatin (Type B) solution was prepared from hides of young bovines, following the method described in [13]. The gelatin solution was dried in a thermostat under air flow, obtaining gelatin solid sheets. The gelatin sheets were freeze dried (Manifold Freeze Dryer, Millrock Technology) and ground to powder (Lab Benchtop Colloid Mill, Sonic Corporation).

2.2. Gelatin Solutions

The gelatin solution investigated in the present work were prepared using the above described gelatin powder, formic acid (98%, Kanto Ltd), acetic acid (66%, Sigma Aldrich), dimethylformamide (Merck) and deionised distilled water. The solutions with 27%, 30% and 33% (w/v) gelatin content, were prepared by dissolving gelatin powder in formic acid and/or the following solvent mixtures: FA:AA (4:1), FA:DMF (4:1) and FA:AA:DMF (4:0.5:0.5). The solutions were stirred at room temperature for 3h and the impurities and inhomogenities were removed using a stainless steel filter.

2.3. Viscosity Measurements

The apparent viscosity, η_a, of the gelatin solutions was measured at room temperature (25±0.1°C), at 3, 6, 9, 12, 24 and 48 hours after preparation, using a rheoviscosimeter type Rheotest 2 (Prüfgerätewerk Medingen) with coaxial cylinders, following ISO 3219/1993 requirements [14]. The results were analyzed according to the power law relationship between the shear stress and the shear deformation rate.

2.4. Gelatin Films

A reference film of 30% (w/v) gelatin was prepared from gelatin powder dissolved in distilled deionised water. Films were prepared from all the investigated gelatin solutions by casting on glass substrates and drying at room temperature for 24h in a thermostat, under air flow. The experiments were performed at room temperature in a room with less then 50% relative humidity.

2.5. Gelatin Electrospinning

The electrospinning set-up consisted of a 10 ml syringe with stainless steel blunt needle (0.5 mm inner diameter), a home-made syringe pump, an aluminium foil as fiber collector, and a Brandenburg Alpha III Series Precision Laboratory HV Power Supply (30V-30kV). The spinning geometry was vertical, the syringe needle being placed at 14 cm above the center of the collector.

The connection between the syringe and the needle was done using a 40 cm long Teflon tube of 0.8 mm inner diameter. The syringe needle was the anode, and the aluminium foil was the cathode. All the solutions were electrospun using the same electrospinning parameters: 19.5 kV dc voltage applied between needle and plate, and 3.7 μL/min solution flow rate.

2.6. Structural Characterization

The morphology of the electrospun gelatin nanofibers was examined by scanning electron microscopy (SEM) using a Vega 2 Tescan (Czech Republic) microscope with Atlas Tescan software for image analysis. The chemical structure of the gelatin films and nanofiber membranes was analyzed by Fourier-transform infrared attenuated total reflectance spectroscopy (FTIR-ATR) using a DIGILAB – SCIMITAR Series FTS 2000 spectrometer with ZnSe crystal, 750-4000 cm^{-1} range, 4 cm^{-1} resolution. The crystallinity of the gelatin nanofibers was evaluated from wide-angle X-ray diffractograms recorded with a Philips X'Pert Pro Multipurpose X-ray Diffractometer operated at 40 kV and 40 mA. Thermal properties of the electrospun membranes and gelatin films were analyzed by recording differential scanning calorimetry (DSC) thermograms of 10 mg samples, using a Thermal Analysis Instrument TA 2910, at a scanning rate of 10°C/min and nitrogen gas flow rate of 50 ml/min.

3. RESULTS AND DISCUSSIONS

3.1. Gelatin Solution Stability

In our experiments we noticed that acetic acid added to a gelatin-FA solution slows down the gelatin degradation process and increases the solution viscosity, while DMF decreases the viscosity and makes the solution stable for a longer time. On the basis of these results, we supposed that adding both AA and DMF to a gelatin-FA solution would produce an electrospinnable solution with improved properties. After a series of tests, we found that a FA:AA:DMF mixture with 4:0.5:0.5 ratio, provides the longest stability and electrospinnability interval (Figure 1). The apparent viscosity dependences on time and gelatin concentration of some solutions investigated in our work are presented in Figures 1. and 2. It is well known that weak acids like acetic acid, determine lyotropic solubilisation [15], while DMF which is a hydrophilic aprotic solvent, is able to bind water and decrease the surface tension of a solution. When acting together, acetic acid and DMF improve gelatin swelling by the two mechanisms: the lyotropic swelling due to the water brought by the non-ionized acid species (water bound by hydrogen bonds to non-ionisable groups), and the osmotic swelling by releasing groups involved in intermolecular interactions through hydrogen bonds [16].

The increased stability of the gelatin solutions prepared with FA:AA:DMF solvent mixture is probably due to the formation of temporary hydrogen bonds and helix-type association of gelatin molecules that keep the solution structurally stable. The formation of gelatin molecule associations intensifies when the gelatin concentration increases, as proved by the viscosity data presented in Fig. 2 for gelatin-FA:AA solutions.

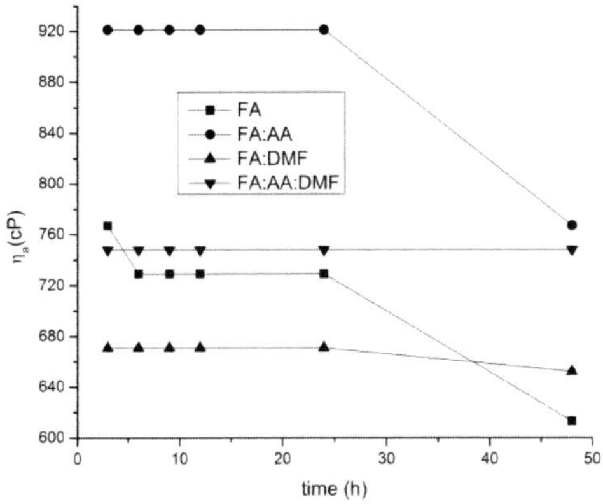

Figure 1. Time dependence of the apparent viscosity, ηa, of gelatin solutions prepared using different solvent mixtures (the lines are just eye guides).

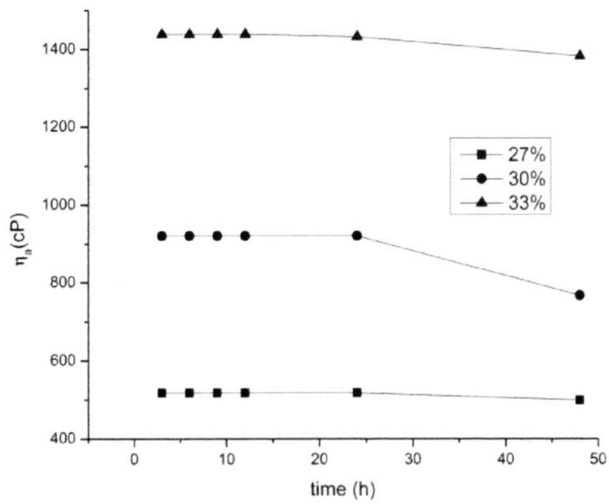

Figure 2. The influence of the gelatin concentration on the apparent viscosity (ηa) time variation of a gelatin-FA:AA solution (percents in the graph legend show the gelatin concentration).

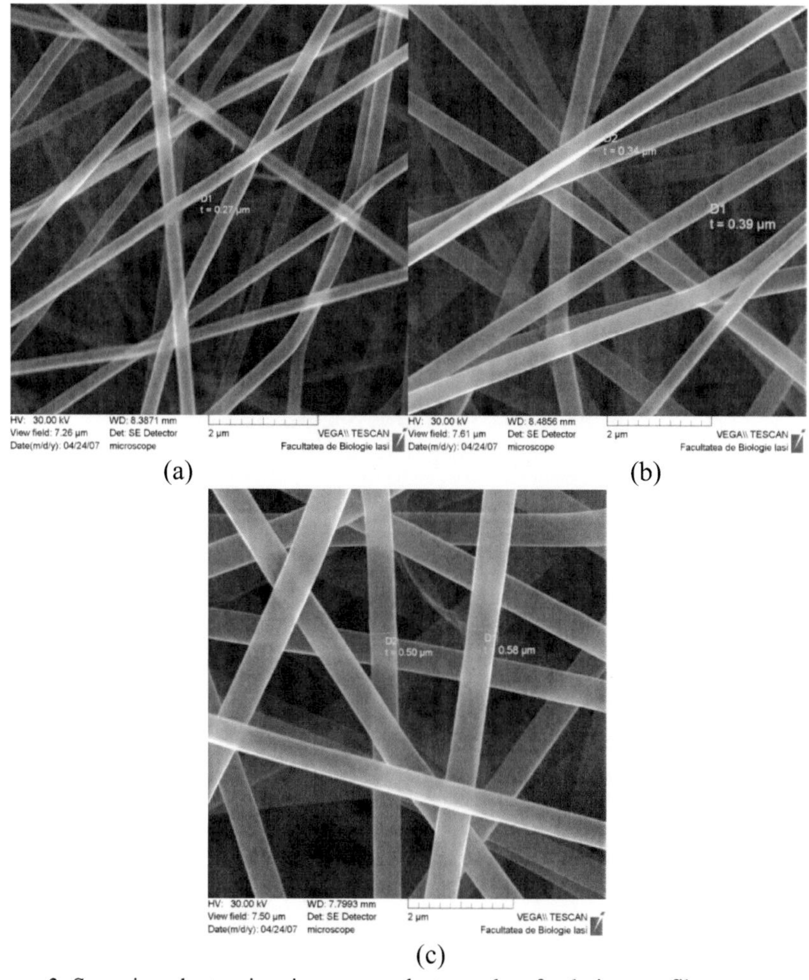

Figure 3. Scanning electronic microscopy photographs of gelatin nanofibers electrospun from gelatin dissolved in formic acid - acetic acid mixture, FA:AA = 4:1 ratio. Gelatin content (w/v): a) 27%, b) 30%, c) 33%.

3.2. Gelatin Nanofiber Morphology

In our electrospinning experiments performed at room temperature, continuous nanofibers were successfully electrospun from all the solutions mentioned above. The nanofibers morphology and the average diameter of the nanofibers were determined from SEM images taken at ten different locations on the surface of each investigated membrane. The gelatin electrospun membranes consisted of uniform nanofibers deposited randomly on the collector surface, without connecting at intersection places, arranged in a 3D structure with pores of about 2 micrometers. Using the same electrospinning parameters mentioned above (i.e. 14 cm, 19.5 kV, 3.7 μL/min), the gelatin-FA:AA:DMF solutions with 27% and 30% (w/v) gelatin, allowed the deposition of sufficiently thick nanofiber membranes in only few minutes (Figure 3). In case of solutions with 33% (w/v) gelatin, we observed the same phenomenon reported by other authors [3]. Periodically, a cloggy particle developed on the edge of the needle tip. When the particle diameter reached about 1 mm, it was detached by the incoming jet and carried towards the collector. This phenomenon was observed for all the solutions of gelatin-FA:AA and gelatin-FA:DMF type. However, because the development of a hardened gelatin particle took about 15 minutes, there was enough time to collect nanofiber membranes without defects for all the gelatin-FA:AA solutions. Increasing the gelatin content of a FA:AA type solution from 27% to 33% (w/v), determined a rapid increase of the nanofibers average diameter from 260 nm to more than double. The SEM images presented in Figure 4. illustrate the effect of various solvent mixtures on the morphology of gelatin nanofibers prepared from solutions of 30% (w/v) gelatin. Figure 4(a) shows nanofibers with 300 nm average diameter, prepared from a gelatin-FA solution. After adding DMF, as known from the viscosity measurements, the apparent viscosity of the solution decreased and the nanofibers became thinner, with an average diameter of 240 nm (Figure 4(b)). When all the solvents acted together in a gelatin-FA:AA:DMF solution, the average diameter of nanofibers increased at 420 nm, in agreement with the increase of the apparent viscosity due to the acetic acid action (Figure 1). The nanofiber average diameters observed in our investigation are presented in Table 1.

Table 1. Gelatin nanofibers average diameter (d_{av}) as a function of solvent type and gelatin concentration (c_G) of the electrospun solution

Solvent	cG (% w)	dav(nm)
FA	30	300
FA:AA	27	260
FA:AA	30	360
FA:AA	33	540
FA:DMF	30	240
FA:AA:DMF	30	420

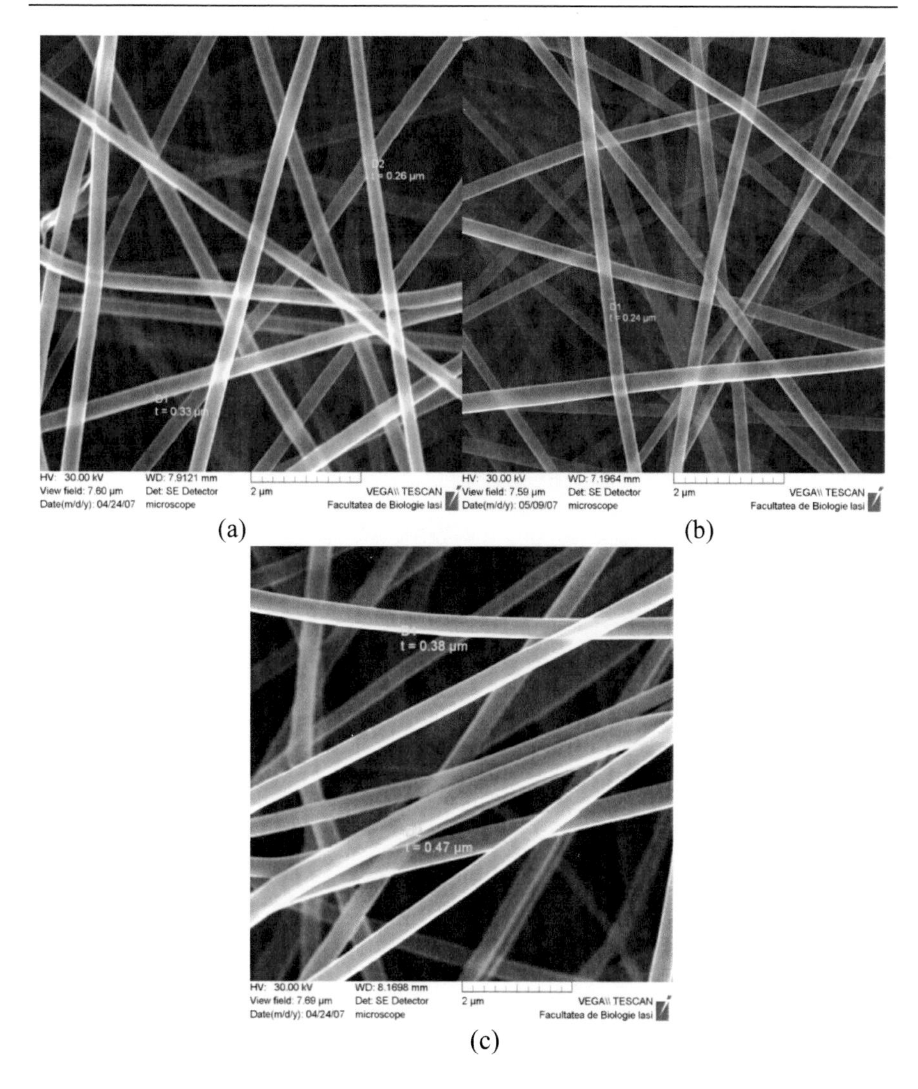

Figure 4. Scanning electronic microscopy images of gelatin nanofibers prepared by electrospinning of gelatin dissolved in: a) FA, b) FA:DMF, c) FA:AA:DMF. The gelatin concentration of all solutions: 30% (w/v).

3.3. Structural Characterization by FTIR, DSC and XRD

Figure 5.(a) shows important differences between the FTIR spectra of gelatin films and those of gelatin nanofiber membranes. The gelatin films prepared with solvent mixtures have peaks at 2922 cm^{-1} and 2848 cm^{-1} (C-H stretching), due to reorientation of gelatin molecules interconnected by CH_2 peptide bonds [17]. These bonds are promoted by the formic acid and do not exist in films prepared from aqueous solutions. The structuring process of gelatin films is favoured by the slow evaporation of solvents. On the contrary, in gelatin nanofibers the evaporation of solvents is very fast and the molecular reorientation is difficult. As a consequence, the two C-H stretching peaks are absent in the FTIR spectra of gelatin nanofibers presented in Fig. 5(b).

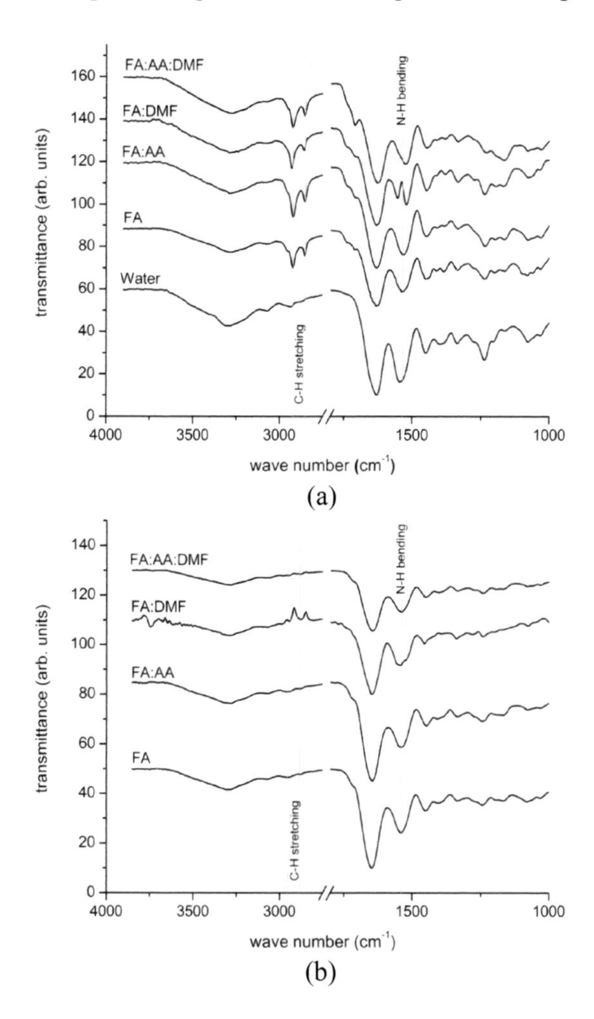

Figure 5. FTIR spectra of gelatin films (a), and electrospun gelatin membranes (b), prepared using various solvent mixtures. Gelatin concentration of all solutions: 30% (w). The two small peaks on the FA:DMF trace near the C-H stretching band are due to the aluminium foil substrate.

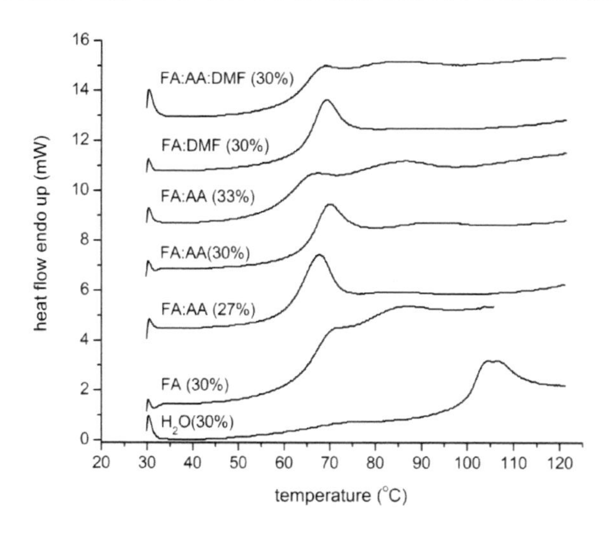

Figure 6. DSC thermograms of gelatin films prepared using various solvent mixtures. The percent in parenthesis shows the gelatin concentration.

The FTIR spectra of gelatin films prepared with FA:DMF mixture show that the N-H bending peak at 1530 cm^{-1} (Amide II) splits into two peaks located at 1560 cm^{-1} and 1515 cm^{-1}. This splitting is probably due to the DMF action on the peptide group, allowing the formation of -CNH links while blocking the formation of carboxyl (-COOH) and amide (-CONH) links. The predominant -CNH links determined the formation of a more ordered structure of gelatin-FA:DMF films during drying, as confirmed by the DSC trace of the gelatin-FA:DMF film characterized by a single endothermic peak (Fig. 6).

The FTIR spectra of the nanofiber membranes (Fig. 5(b)) look almost identical, irrespective of the solvent mixture used. This is a good result, proving that there is no need for further treatments to remove undesired structures in the electrospun gelatin nanofiber mats. The identical structure of the electrospun gelatin nanofibers is a result of the intense electric field action and of the very fast evaporation of solvents from the solution filaments. The rapid elongation of the solution jet stretched by electrical forces breaks the interchain bonds, while the fast evaporation of solvents freezes the disordered state leading to the formation of amorphous nanofibers.

The amorphous structure of the electrospun gelatin nanofibers is confirmed by the X-ray diffractograms presented in Fig. 7. The broad peaks on some of the X-ray diffractograms are coming from the different fixtures used to hold the investigated samples. We used various methods in searching for some preferential molecular orientation in the gelatin nanofibers, but all the membranes appeared completely amorphous. The different molecular ordering processes promoted in gelatin films by the various solvent mixtures are clearly visible on the DSC thermograms (Figure 6). The gelatin film prepared form an aqueous solution shows a main endothermic peak at 125°C, due to the glass transition superposed with water evaporation. On the contrary, almost all the gelatin films prepared with solvent mixtures show two endothermic peaks, around 68°C and 88°C. The difference is made, the same as in case of FTIR spectra, by the film prepared with the FA:DMF mixture which shows only one endothermic peak at 69.5°C. The absence of the second endothermic peak is a result of the same molecular interactions that caused Amide II peak splitting observed on the FTIR spectra (Figure 5.(a)), and determined the largest decrease of the viscosity noticed among all the solutions investigated (Figure 1). The second endothermic peak showed by the DSC records of gelatin films presented in Figure 6, appears to be very sensitive to the gelatin concentration. The films prepared from gelatin-FA:AA solutions of 27%, 30% and 33% (w/v) gelatin, show clearly that a higher gelatin concentration increases the surface of the second endothermic peak. This observed increase is probably a consequence of the competition between acetic acid molecules and gelatin molecules to occupy the valence sites on CO and NH groups of the peptide bonds, resulting in different molecular reorientation processes during films drying [18].

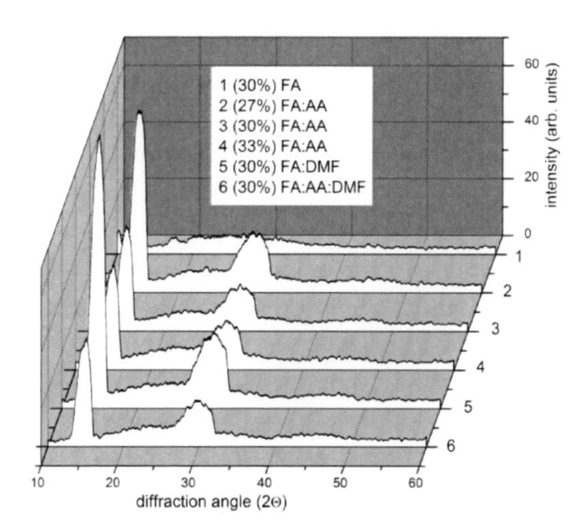

Figure 7. X-ray diffractograms of gelatin nanofiber membranes, showing complete amorphous structure irrespective of the solvent mixture used. The percents on the graph legend refer to the gelatin concentration. The wide peaks at 15° and 28° are coming from the sample holder.

Conclusions

Gelatin nanofibers were successfully prepared by the electrospinning of gelatin solutions with gelatin concentrations of 27% and 30% (w/v). The membranes consisted of uniform nanofibers randomly oriented, forming 3D structures with pores of about 2 micrometers. Increasing the gelatin concentration caused the increase of the gelatin nanofiber diameter from 200 nm to 580 nm.

Various mixtures of formic acid, acetic acid and dimethylformamide were used in searching for an optimized gelatin solution with maximum gelatin concentration and longest stability. It was found that a solvent mixture of FA:AA:DMF with 4:0.5:0.5 ratio gave the best results, the solution remaining stable and spinnable for more than 48 hours. The highest gelatin concentration of a FA:AA:DMF type solution allowing a good quality and stable electrospinning process was, in our experiments, 30% (w/v).

Electrospinning of gelatin solutions prepared with a FA:AA:DMF solvent mixture does not require adjustment of the electrospinning parameters when the gelatin concentration varies in the range 27-30% (w/v), nanofiber membranes being successfully prepared from these solutions using the same electrospinning parameters.

Electrospinning of high concentration gelatin FA:AA:DMF solutions allows a fast deposition of gelatin non-woven mats of enough thickness and mechanical strength required in biomedical applications.

References

[1] M. Li, M. J. Mondrinos, M. R. Gandhi, F. K. Ko, A. S. Weiss, P. I. Lelkes, *Biomaterials* 26, 5999 (2005).

[2] S. Liao, B. Li, Z. Ma, H. Wei, C. Chan and S. Ramakrishna, *Biomed Mater* 1, R45 (2006).

[3] Z. M. Huang, Y. Z. Zhang, S. Ramakrishna, C. T. Lim, *Polymer* 45, 5361 (2004).

[4] R. A. Hule, D. J. Pochan, *MRS Bulletin* 32, 354 (2007).

[5] C. G. B. Cole, Gelatin in Encyclopedia of Food Science and Technology, 2nd edition, *John Wiley and Sons* (2000).

[6] F. H. C. Crick, A. Rich, *The Structure of Collagen in Recent Advances in Gelatin and Glue Research*, Pergamon Press (1958).

[7] A. Courts, *Biochem. J.* 83, 124 (1962).
[8] P. V. Kozlov, G. I. Burdygina, *Polymer* 24, 651 (1983).
[9] K. Gast, A. Siemer, D. Zirwer, G. Damaschun, *Eur. Biophys. J.* 30, 273 (2001).
[10] C. S. Ki, D. H. Baek, K. D. Gang, K. H. Lee, I. C. Um, Y. H. Park, *Polymer* 46, 5094 (2005).
[11] R. V. D. Montenegro, Dissertation, Max-Planck-Institut für Kolloid- und Grenzflächenforschung (2003).
[12] K. Landfester, Preparation of polymer and hybrid colloids by miniemulsion for biomedical applications, in Colloid Polymers: preparation and biomedical applications, Ed. A. Elaïssari, Dekker, New York/Basel (2003).
[13] C. G. B. Cole, J. J. Roberts, *The SA Journal of Food Science and Nutrition* 8, 139 (1996).
[14] ISO 3219:1993. Plastics – Polymers/resins in the liquid state or as emulsions or dispersions –Determination of viscosity using rotational viscometer with defined shear rate.
[15] C. M. Ofner 3rd, H. Schott, *J. Pharm. Sci.* 75(8), 790 (1986).
[16] C. M. Ofner 3rd, H. Schott, *J. Pharm. Sci.* 76(9), 715 (1987).
[17] D. A. Prystupa, A. M. Donald, *Polymer Gels and Networks* 4, 87 (1996).
[18] K. H. Gustavson, *The Chemistry and Reactivity of Collagen*, Academic Press (1956).

In: Electrospinning Process and Nanofiber… ISBN 978-1-61209-330-7
Editors: A.K. Haghi and G.E. Zaikov

Chapter 3

ELECTROSPUN GELATIN NANOFIBERS FUNCTIONALIZED WITH SILVER NANOPARTICLES

Florentina Tofoleanu[1], Tudorel Balau Mindru[2], Florin Brinza[1], Nicolae Sulitanu[1], Ioan-Gabriel Sandu[3], Dan Raileanu[4], Viorel Floristean[5], Bogdan Alexandru Hagiu[5], Cezar Ionescu[6], Ion Sandu[7] and Vasile Tura[*1]

[1]Faculty of Physics, Al. I. Cuza University, Blvd. Carol I nr. 11A, 700506, Iasi, Romania
[2]Faculty of Textiles and Leather Engineering, Gh. Asachi Technical University, Str. Dimitrie Mangeron nr. 53, Corp TEX1, 700050, Iasi, Romania
[3]Faculty of Materials Science and Engineering, Gh. Asachi Technical University, Str. Dimitrie Mangeron nr. 63, 700050, Iasi, Romania
[4]Faculty of Biology, Al. I. Cuza University, Blvd. Carol I nr. 11A, 700506, Iasi, Romania
[5]Faculty of Veterinary Medicine, Ion Ionescu de la Brad University of Agricultural Science and Veterinary Medicine, Aleea Mihail Sadoveanu nr. 3, Iasi, 700490, Romania
[6]Faculty of Medicine, Gr. T. Popa University of Medicine and Pharmacy, Str.Universitatii nr.16, 700115, Iasi, Romania
[7]Faculty of Orthodox Theology, Al. I. Cuza University of Iasi, Bd. Carol I, nr. 11A, 700506, Romania

ABSTRACT

The present chapter deals with gelatin nanofibres functionalized with silver nanoparticles, prepared by electrospinning using solutions of gelatin mixed with silver nitrate. As a common solvent for gelatin and $AgNO_3$ was selected a mixture of formic acid and acetic acid in volume ratio 4:1. In this system, formic acid was used as a solvent of gelatine, but also as reducing agent for silver ions in solution. Silver nanoparticles were stabilized through a mechanism that involves an interaction with oxygen atoms of carbonyl groups of gelatin. The gelatin nanofibres functionalised with silver nanoparticles were characterized by transmission electron microscopy (TEM), scanning electron microscopy (SEM), X-ray diffraction (XRD) and antimicrobial test. The results of investigations by TEM and XRD confirmed the presence of silver nanoparticles with diameters less than 20 nm, uniformly distributed over the surface of smooth nanofibres with an average diameter of 70 nm. The tests demonstrated that gelatin/Ag nanofibers have a good antimicrobial activity against *Escherichia coli*.

Keywords: Electrospinning; Gelatin; Silver; Nanofibers; Nanoparticles.

1. INTRODUCTION

One of the very important tissues for human health is the skin, which plays the role of an interface between body and environment, and acts as a protective barrier against physical, chemical and biological aggressions. The skin is actually a complex system involved in the processes of absorption and excretion, showing a selective permeability for some important substances in human metabolism [1].

The currently increasing incidence of skin diseases caused a significant demand for skin replacement materials which in many cases can not be assured from natural sources. It also increased the demand for skin regrowth substrates used in treating skin lesions, which stimulated tissue engineering to develop new biodegradable materials which can promote adhesion, development and migration of fibroblasts, necessary for a good quality reepithelization [2]. Skin substitutes used today are expensive and have limitations concerning the biocompatibility, making their use sometimes risky [3].

There have been developed various technologies to produce biomaterials used for skin lesions treatment of wound dressings. One of the very interesting nanotechnologies investigated now is polymer solutions spinning in electrostatic field (electrospinning), which can be applied to solutions of various natural and synthetic polymers such as gelatine, collagen, chitosan, fibrinogen, silk, polylactic acid, hyaluronic acid, obtaining non-woven nanofibres membrane with thicknesses in a range that goes from a few tens of nanometers to over one thousand nanometers [4-6]. This type of three-dimensional structure of nanofibres prepared by electrospinning resembles very well with the human extracellular matrix, which explains the great interest manifested today for this nano-technology.

The biopolymers have high viscosity and very low solubility in common organic solvents, which makes them difficult to process by solution spinning in electrostatic field. Natural polymers dissolve well in 1,1,1,3,3,3-hexafluoro-2-propanol and trifluoroacetic acid [7], which are toxic and expensive solvents, features that limit their use in applications on an industrial scale [8-11].

Gelatin is a biopolymer widely used in pharmaceutical industry and for medical device applications. As a result of the growing interest for regenerative medicine, in the recent past gelatin has begun to be studied also as a candidate material for obtaining scaffolds with similar qualities of natural media for a wide range of tissues [12-16].

In a previous study we reported the preparation of nanofibres with superior mechanical and biological characteristics, by electrospinning of gelatin solutions with concentrations up to 30% (w/v) [17]. In the present paper we investigated the possibility to improve the properties and extend the applicability of gelatin nanofibers by adding silver nanoparticles which are known to have a positive effect in wounds healing and skin regeneration.

2. Experimental

2.1. Gelatin/$AgNO_3$ Solutions

The electrospinnable solutions used in our experiments were prepared using gelatin (Fisher HealthCare, MP Biomedicals), formic acid (98 +100%, Scharlau Chemie SA), acetic acid (99.5%, Chemical Company SA), silver nitrate (99.5% Chemical Company SA) and distilled deionised water. The reagents were used as received from the manufacturer.

In the beginning, a solution of 27% (w/v) gelatin in a mixture of formic acid with acetic acid in the volumetric ratio 4:1 was prepared. The solution was subjected to stirring at room temperature for 3h, after which the impurities and undissolved parts were eliminated using a stainless steel filter.

The silver precursor solution was prepared by dissolving 5% (w/v) $AgNO_3$ in 66% (v/v) acetic acid in distilled deionised water.

The electrospinnable solution was prepared by adding dropwise the $AgNO_3$ solution to the gelatin solution, under continuous stirring for 20 minutes.

The solutions viscosity was determined at room temperature using a Nahita Rotary Viscometer, observing the rules ISO 3219/1993 [18].

2.2. Electrospinning

The equipment for spinning in electrostatic field was made of a 10 ml syringe filled with solution, actuated by a home-made syringe pump able to provide flow rates in the range 1.0-40.0 μL/min. The syringe was connected through a 40 cm Teflon tube of 0.5 mm inner diameter to a blunt needle placed 12 cm high above a collector covered with aluminium foil. Between the needle and collector high-voltages in the range 11.0-19.5 kV have been applied using a Brandenburg Alpha III Series HV Power Supply (30V-30kV). The optimum solution flow-rate for obtaining good quality nanofibers was 2.4μL/min.

2.3. Crosslinking with Glutaraldehyde

The water stability of the nanofibers was improved by crosslinking with glutaraldehyde vapours (25% solution) for 24 h, drying at room temperature for 2 h and heating in a thermostat at 70°C for 2h. It is known that such a treatment promotes gelatin chains crosslinking by formation of Schiff bases type iminic groups [19,20].

2.3. Structural Characterization

The morphology of the electrospun gelatin nanofibers was examined by scanning electron microscopy (SEM) using a Vega 2 Tescan (Czech Republic) microscope. The nanofibers diameter was estimated using Atlas Tescan software for image analysis.

The chemical structure of nanofibers was investigated by Fourier-transform infrared attenuated total reflectance spectroscopy (FTIR-ATR) using a DIGILAB – SCIMITAR Series FTS 2000 spectrometer with ZnSe crystal, working in the range 750-4000 cm^{-1}, with 4 cm^{-1} resolution.

Dimensional analysis and structural characterization of silver nanoparticles was carried out by transmission electron microscopy (TEM) using a Philips CM100 microscope, and by X-rays diffraction (XRD) using a DRON 2.0 diffractometer with Cu tube (Kα radiation).

2.4. Antimicrobial Activity Characterization

The antibacterial activity was tested on solid media by the disk diffusion method. The bacterium inocula were prepared from a culture of *Escherichia coli* ATCC 25922 overnight incubated at 37°C in nutrient broth, by diluting with peptoned water to a 0.5 McFarland standard. Petri plates containing Mueller Hinton agar were innoculated by inundation with prepared microbial suspension.

Two samples of the investigated material, one containing Ag nanoparticles and one without Ag nanoparticles (used as control) were placed on the surface of the inoculated media. The plates were incubated at 37°C for 24 hours. Following incubation, plates were examined in order to identify zones of no growth (halos around the fragments) characteristic for antimicrobial activity.

3. Results and Discussions

In case of electrospinning gelatine, its strong polar nature makes it difficult to solve in common solvents. The special solvents mentioned above are toxic and very expensive.

Using the results of our previous study, in the present work we choosed as gelatin solvent a mixture of formic acid and acetic acid in volumetric ratio 4:1 that provided a better solution stability and a longer electrospinnability interval [17], due to the acetic acid which determined a dynamic rebuild of the hydrogen bonds involved in generating helicoidal structures of gelatin molecules.

In the present work, by electrospinning a solution of AgNO3 in gelatin, good quality nanofibers with smooth surface and without beads were obtained, as proved by the SEM images presented in Figures 1, 2.

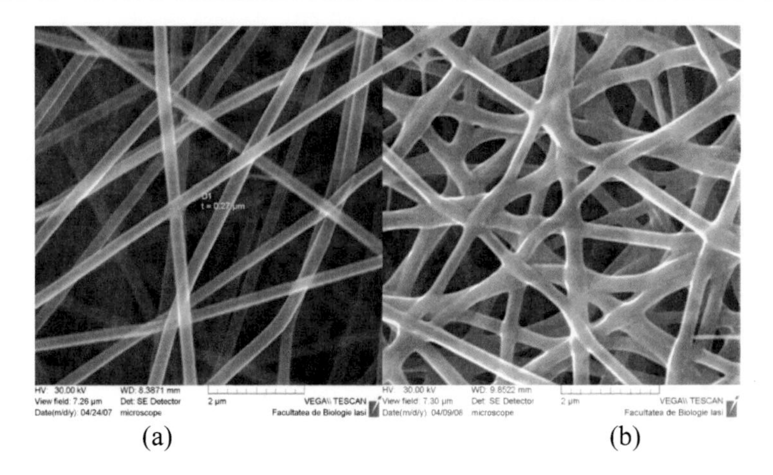

Figure 1. SEM images of gelatin nanofibres prepared by electrospinning a solution of gelatin in formic acid and acetic acid: (a) as electrospun, (b)crosslinked in vapours of glutaraldehyde.

The morphology of the nanofibers shown in Figure 2.(a) suggest that adding 5% (w/w) silver nitrate to gelatin resulted in an increased solution conductivity and caused a slight decrease of fibers diameter. Moreover, the increased conductivity and decreased viscosity of the electrospun solution determined occasionally the formation of some auxiliary microjets which formed nanofibres with diameters less than 30 nm, intercalated among majority nanofibres with an average diameter of 190 nm.

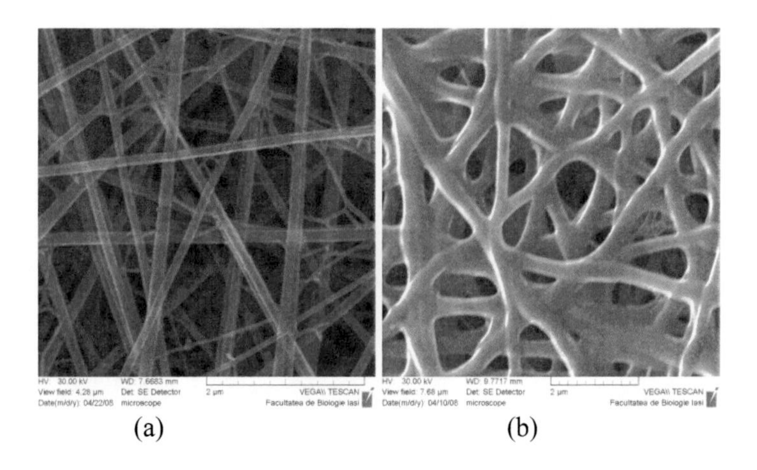

Figure 2. SEM images of gelatin nanofibers functionalised with silver nanoparticles: (a) as electrospun, (b) crosslinked in vapours of glutaraldehyde.

$$\text{gelatin-NH}_2 + \text{HOC}\text{~~~}\text{COH} + \text{NH}_2\text{-gelatin}$$

$$\Downarrow$$

$$\text{gelatin-N}{=}\text{C}\text{~~~}\text{C}{=}\text{N-gelatin} + 2H_2O$$

Further treatment of gelatin or gelatin/Ag nanofibers with glutaraldehye vapours resulted in a slight increase of nanofibers diameter due to water removal and increase of material density (Figures 1.b, 2.b). The water removed during the crosslinking process accumulated on the nanofibers surface and caused a superficial dissolution and bonding of nanofibers at the contact points.

As compared with pure gelatin nanofibers in Figure 1.b, the crosslinked gelatin/Ag nanofibers shown in Figure 2.b. appear to be more affected by water, probably because an increased hydrophylicity as a consequence of structural changes induced by silver nanoparticles surfaces. Images obtained by TEM (Figure 3) show the presence of some nanoparticles on the surface and in the volume of gelatin nanofibers. The shape of nanoparticles is spherical, with diameters less than 20 nm.

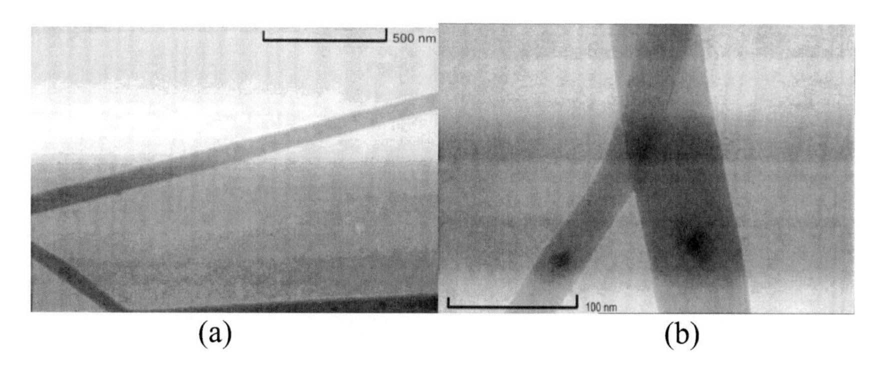

(a) (b)

Figure 3. TEM images of silver nanoparticles on the surface (a), and inside (b) *of electrospun gelatin nanofibers.*

3.3. Structural Characterization by FTIR and XRD

The infrared spectra showed in Figure 4 show that the spectrum of gelatine nanofibers is almost identical with the spectrum of gelatin/Ag nanofibers, with just two small differences. First, it can be observed a more intense absorption in the band at 1406 cm^{-1}, assigned to symmetrical stretching of the carboxyl group, -COOH. Second, the absorption peaks at 2922 cm^{-1} and 2848 cm^{-1}, suggest that the silver nanoparticles caused a reordering of gelatin chains by promoting the formation of peptide CH_2 links [21]. This restructuring effect is very similar to the effect reported in our previous paper [17], where the bands at 2922 cm^{-1} and 2848 cm^{-1} appeared in infrared spectra of films prepared from aqueous gelatin solutions and were not present in spectra of gelatin nanofibers. The high relative elongation of electrospun nanofibers and the fast solvent evaporation freezes an amorphous structure in pure gelatin nanofibers. By contrast, as the spectra from Figure 4 prove, the presence of silver nanoparticles stimulates the restoration of local order of molecular chains and facilitates the formation of peptide links between them.

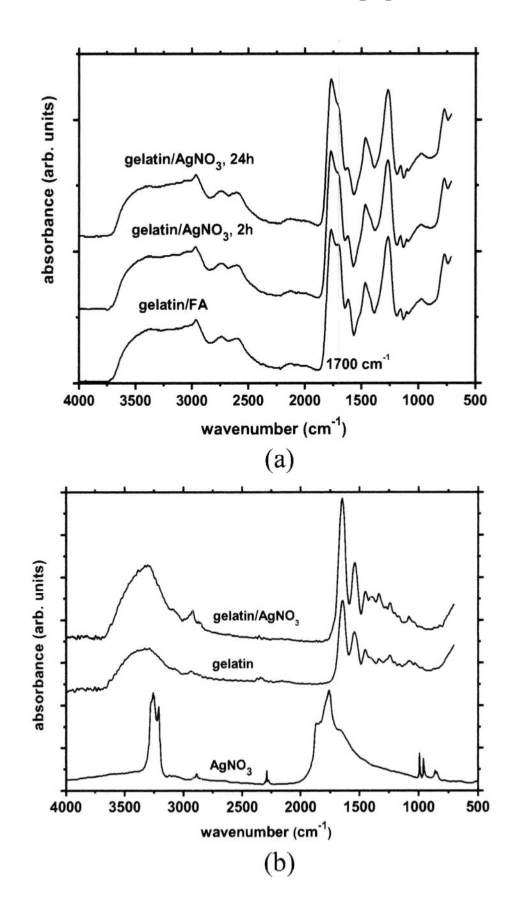

Figure 4. FT-IR spectra: (a) gelatin/AgNO3 solution, (b) electrospun gelatin nanofibers doped with silver nanoparticles.

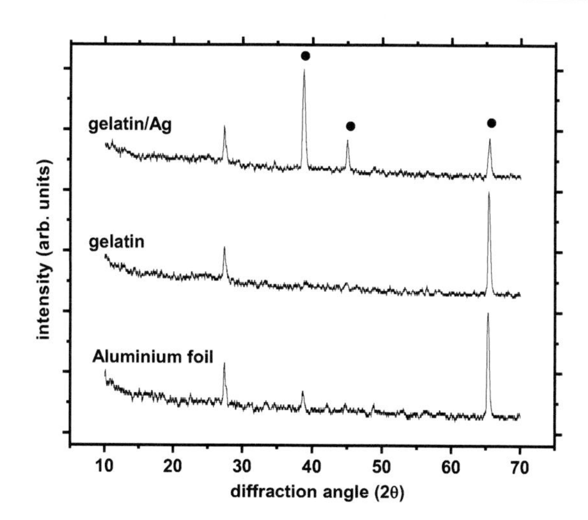

Figure 5. X-ray diffractograms of electrospun gelatin nanofibers functionalised with silver nanoparticles, crosslinked with glutaraldehyde.

In the FTIR spectrum of the gelatin/$AgNO_3$ solution in Figure 4.a, one can see a slight increase in the intensity of the absorption bands amide I and II, but without any new peaks that could be assigned to a possible unreacted phase of $AgNO_3$. On the basis of Figure 4.a. it can be assumed that the entire amount of silver nitrate reacted with the formic acid and silver is present in the gelatin/Ag nanofibers only in metal form. The presence of silver only in the form of well crystallised metal nanoparticles is certified by the X-ray diffractogramme shown in Figure 5, where the peaks due metallic silver are clearly present.

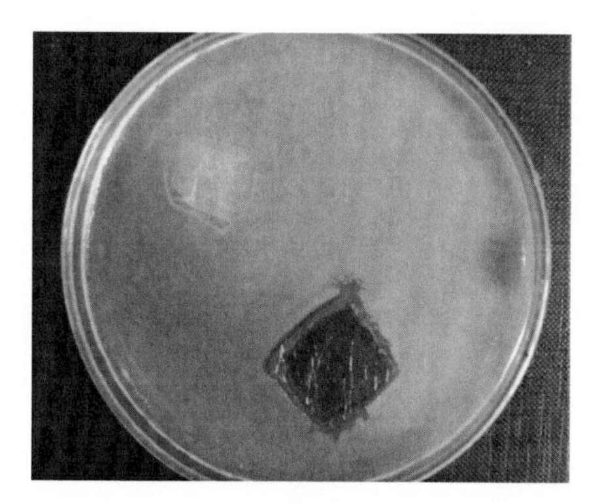

Figure 6. Antimicrobial activity of gelatin nanofibers against Escherichia coli ATCC 25922. Upper left – gelatin, bottom right - gelatin with silver nanoparticles.

CONCLUSION

In the present work, there were prepared non-woven membranes made of gelatin nanofibers functionalized with silver nanoparticles with diameters less than 20 nm. After crosslinking with glutaraldehyde vapours, the nanofibers gained good water stability.

The structural analysis by infrared spectroscopy and X-ray diffractometry have shown that in the final nanofibers silver is present only in the form of well crystallised metal nanoparticles.

The presence of growth inhibition zones, i.e. halos around the samples of gelatin/Ag nanofibers proved the antimicrobial activity of the investigated material against *Escherichia coli* ATCC 25922.

The superior physical and chemical characteristics, along with a good biocompatibility and other improved properties as a lack of inflammatory reaction and relatively fast resorbtion, recommend the gelatin nanofibers functionalized with Ag nanoparticles as a good candidate material for the manufacture of wound dressings skin tissue scaffolds.

REFERENCES

[1] D. N. Shier, J. L. Butler, R. Lewis, Hole's Human Anatomy and Physiology, 11th edition, McGraw-Hill Higher Education (2006).

[2] R. Langer, J. P.Vacanti, *Science* 260, 920 (1993).

[3] G. Damodaran, M. Syed, I. Leigh, S. Myers and H. Navsaria, *Expert Review of Dermatology* 3(3), 345 (2008).

[4] X. Wen, D. Shi, N. Zhang, Applications of Nanotechnology in Tissue Engineering, in Handbook of Nanostructured Biomaterials and Their Applications in Nanobiotechnology, Ed. H. S. Nalwa, American Scientific Publishers, 1 (2005).

[5] Y. Zhang, H. Ouyang, C. T. Lim, S. Ramarkrishina and Z. M. Huang, *J. Biomed. Mater. Res. Part B: Appl. Biomater.* 72B, 156 (2005).

[6] H. W. Kim, J. H. Song and H. E. Kim, Adv. *Funct. Mater.* 15, 1988, (2005).

[7] R. Murugan, S. Ramakrishna, *Tissue Engineering* 12(3), 435 (2006).

[8] M. Li, M. J. Mondrinos, M. R. Gandhi, F. K. Ko, A. S. Weiss and P. I. Lelkes, *Biomaterials* 26, 5999 (2005).

[9] G. E. Wnek, M. E. Carr, D. G. Simpson and G. L. Bowlin, *Nano Lett.* 3, 213 (2003).
[10] K. Ohkawa, D. Cha, H. Kim, A. Nishida and H. Yamamoto, *Macromol. Rapid Commun.* 25, 1600 (2004).
[11] J. A. Matthews, G. E. Wnek, D. G. Simpson and G. L. Bowlin, *Biomacromolecules* 3, 232 (2002).
[12] M. Li, M. J. Mondrinos, M. R. Gandhi, F. K. Ko, A. S. Weiss, P. I. Lelkes, *Biomaterials* 26 (2005) 5999-6008.
[13] S. Liao, L. Bojun, M. Zuwei, H. Wei, C. Chan and S. Ramakrishna, *Biomed. Mater.* 1, R45 (2006).
[14] R. Langer and D. A. Tirrell, *Nature* 428, 487 (2004).
[15] P. Y. W. Dankers, M. C. Harmsen, L. A. Brouwer, M. J. van Luyn and E. W. Meijer, *Nature Mater.* 4, 568 (2005).
[16] Z. M. Huang, Y. Z. Zhang, M. Kotaki and S. Ramarkrishna, *Compos. Sci. Technol.* 63, 2223 (2005).
[17] T. Balau Mindru, I. Balau Mindru, T. Malutan, V. Tura, *Journal of Optoelectronics and Advanced Materials*, 9(11), 3633 (2007).
[18] ISO 3219:1993. Plastics - Polymers/resins in the liquid state or as emulsions or dispersions - Determination of viscosity using rotational viscometer with defined shear rate.
[19] Y. Koyama, A. Taniguchi, *J. Appl. Polym. Sci.*, 31(6), 1951 (1986).
[20] C. Tual, E. Espuche, M. Escoubes, A. J. Domard, *Polym. Sci., Part B: Polym. Phys.*, 38(11), 1521 (2000).
[21] D. A. Prystupa and A. M. Donald, *Polymer Gels and Networks* 4, 87 (1996).

In: Electrospinning Process and Nanofiber... ISBN 978-1-61209-330-7
Editors: A.K. Haghi and G.E. Zaikov

Chapter 4

ELECTROSPINNING OF GELATIN/CHITIN COMPOSITE NANOFIBERS

Vasile Tura[1], Florentina Tofoleanu[1], Ionel Mangalagiu[2], Tudorel Balau Mindru[3], Florin Brinza[1], Nicolae Sulitanu[1], Ion Sandu[4], Dan Raileanu[5] and Cezar Ionescu[6]

[1]Faculty of Physics, Al. I. Cuza University, Blvd. Carol I, nr. 11A, 700506, Iasi, Romania
[2]Faculty of Chemistry, Al. I. Cuza University, Blvd. Carol I, nr. 11A, 700506, Iasi, Romania
[3]Faculty of Textiles and Leather Engineering, Gh. Asachi Technical University, Str. Dimitrie Mangeron nr. 53, Corp TEX1 700050, Iasi, Romania
[4]Faculty of Orthodox Theology, Al. I. Cuza University of Iasi, Blvd Carol I, nr. 11A, 700506, Romania
[5]Faculty of Biology, Al. I. Cuza University, Blvd. Carol I, nr. 11A, 700506, Iasi, Romania
[6]Faculty of Medicine, Gr. T. Popa University of Medicine and Pharmacy,Str.Universitatii nr.16, 700115, Iasi, Romania

ABSTRACT

Gelatin/chitin nanofibers were prepared by electrospinning. Various solutions were obtained by mixing a 27 % (w/v) gelatin in formic acid solution with crab shell chitin dissolved in a mixture of trichloroacetic acid and formic acid in 1:4 volume ratios.

Obtaining electrospinnable gelatin/chitin solutions required the decrease of chitin molecular mass. Two types of depolymerisation processes were tested: microwave irradiation and ultrasonic treatment.

By variation of irradiation parameters, an optimum was found with the microwave reactor working in temperature control mode. The chemical and physical structure of the gelatin/chitin nanofibers were investigated by scanning electron microscopy and infrared spectroscopy.

Keywords: Nanofibers; Electrospinning; Gelatin;Chitin; Blend.

1. Introduction

In the last ten years, there was an increasing interest in tissue engineering for the development of scaffolds using biodegradable and biocompatible natural polymers.

In natural tissues, the wound healing process involves cells attachment and proliferation inside the extracellular matrix, a three-dimensional network structure made of collagen multifibrils and proteoglycans, with diameters in the range from 50 nm to 500 nm [1].

A good scaffold should mimic the structure and biological function of the proteins in the natural extracellular matrix, in order to provide mechanical support for cell attachment and facilitate all the necessary cell activities. The scaffold material and its texture were reported to have an important influence on the shape and function of the regrowth tissue [2-4].

The above mentioned requirements are very well fulfilled by an electrospun nonwoven membrane consisting of nanofibers arranged in a geometry similar to the collagen structure of the natural extracellular matrix. When sugar- and amino- acid residues are included in the structure of these nanofibers, they are able to stimulate physiological responses similar to those triggered by the proteins and growth factors acting in the human extracellular matrix during tissue regeneration and restructuration. In the present work, we investigated a blend of chitin and gelatin as a candidate material for wound dressings. The chitin monomer unit, N-acetylglucosamine, occurs also in hyaluronic acid known as an extracellular macromolecule important in the wound repair processes.

Chitin has also structural characteristics similar to glycosaminoglycans such as chondroitin sulfates which contribute to control the tensile strength of tissues [5].

Gelatin is basically denaturated collagen and offers all the surface structural elements needed for cell attachment and migration. The chitin existing in the gelatin/chitin nanofibers applied on the wound surface could stimulate macrophages for secretion of lysozyme and human chitinase, enabling in this way the immune system to prevent infection by chitin-based pathogens [6].

Besides this, chitin is hydrolyzed to chitooligomers which stimulate other macrophage functions like collagen deposition, nitric oxide and tumor necrosis factor production [7].

Therefore, we believe that nanofibers made of a gelatin/chitin blends may possess all the important characteristics needed for promoting rapid dermal regeneration and accelerating wound healing.

2. Experimental

2.1. Materials

The gelatin/chitin blends investigated in this work were prepared from crab shell chitin (Sigma-Aldrich, practical grade) and gelatin (Fisher, HealthCare), used as received. As solvent for chitin, a mixture of trichloroacetic acid (Scharlau Chemie, reagent grade) and formic acid (Scharlau Chemie, 98-100%) in volumetric ratio 1:4 was used. Gelatin was dissolved in formic acid, 27% (w/v), by stirring at room temperature.

2.2. Chitin Depolymerisation

Chitin solutions were prepared by dispersing the chitin powder in suitable quantities of solvent (trichloroacetic and formic acids), followed by ultrasonic or microwave irradiation. Ultrasonic treatments were performed using a Sonoplus HD 2200 (Bandelin Electronic) ultrasonic homogenizer working at 20 kHz, able to provide sonic power up to 200W, to solution volumes in the range 1-200 ml. For microwave irradiation, a STAR-2 (CHEM Corp.) microwave reactor was used, working at 2.45 GHz frequency and 800 W maximum power.

2.3. Structural Characterization

The apparent viscosity, η_a, of chitin solutions in alkali measured at room temperature (25±0.1°C), using a Nahita Rotary Viscometer 801, observing ISO 3219/1993 requirements [8]. The intrinsic viscosity η of the chitin solution samples dissolved in alkali (2.77 M NaOH) was measured with an Ubbelohde U-tube viscometer using the relation:

$$\eta = t_{ch} / t_s$$

where t_{ch} and t_s are the flowing intervals of the chitin solution and the solvent.

The molecular mass of chitin in all the investigated solutions was computed using the correlation between intrinsic viscosity and molecular weight given by the Mark-Houwink-Sakurada-equation:

$$\eta = KM_W^a$$

where K and the exponent a are temperature dependent parameters for a given polysaccharide solvent system [9, 10]. The exponent a is a polymer conformation parameter that decreases with increasing molecular compactness. In our experiments, the molecular weight of the chitin samples were computed using K=0.10 and a=0.68 [11]. The viscosity measurements on as received chitin gave η=935 ml/g and M_W=1125 kg/mol.

The structural changes of chitin solutions after ultrasonic or microwave treatments were investigated by infrared spectroscopy, using a Bruker Tensor 27 FT-IR Spectrometer with Platinum ATR single reflection diamond ATR sampling module (Bruker Optics), working from 4000 to 400 cm^{-1}.

The chemical structure of the gelatin/chitin nanofiber membranes was analyzed by FTIR-ATR using a DIGILAB – SCIMITAR Series FTS 2000 spectrometer with ZnSe crystal, working in the range 4000-750 cm^{-1} with 4 cm^{-1} resolution.

2.4. Electrospinning

The electrospinning set-up consisted of a 10 ml syringe with stainless steel blunt needle (0.5 mm inner diameter), a home-made syringe pump, an aluminium foil as fiber collector, and a Brandenburg Alpha III Series HV Power Supply (30V-30kV). A vertical geometry was selected in which the syringe needle was placed at 12 cm above the collector center. The syringe and the blunt needle were connected by a 40 cm long Teflon tube of 0.5 mm inner diameter. The syringe needle acted as anode, and the aluminium foil was the cathode. Between needle and collector dc voltages in the range 10-14.5 kV have been applied. The solutions were electrospun at solution flow rates between 2.4 and 3.7 μL/min.

3. Results and Discussions

The most important step in electrospinning a polymer is the preparation of a suitable solution. The electrospinning process stability and the nanofiber morphology are influenced by the physical properties of this solution. An electrically stretched polymer solution jet will form a fiber only if a balance between the surface tension and the viscosity of the solution is acquired. The solution surface tension is very much influenced by the surface tension of the solvent. The solution viscosity depends on the molecular mass of the polymer, which is proportional with the macromolecules length. When polymer molecules are too short, there is insufficient entanglement between the molecular chains and the solvent surface tension becomes dominant, giving rise to beads along the fiber. If polymer molecules are too long, i e. molecular mass too high, the solution viscosity is high and jet stretching by electrical forces becomes difficult or even impossible [12].

The chitin extracted from crab shells (α-chitin) is characterized by a strong three-dimensional hydrogen bond network of antiparallel unit cell (*N*-acetylglucosamine) stacks, which makes swelling and dissolution processes very difficult. As extracted α-chitin from crab shells has a high molecular mass of 1125 kg/ mol, as derived from viscosity measurements.

The poor solubility of chitin is a result of the close packing of chains and its strong intra- and intermolecular bonds between the hydroxyl and acetamide groups [13]. The inability of α-chitin to swell upon soaking in water is explained by the extensive intermolecular hydrogen bonding [14].

Under these circumstances, preparation of an electrospinnable solution requires a supplementary energy that could help the solvent to tear apart and break the long molecular chains. In our investigation, two possibilities of delivering a supplementary energy to the solvent have been tested, ultrasound and microwave (MW) irradiation.

3.1. Chitin Depolymerisation by Ultrasonic Treatment

A first set of experiments was aimed to prepare solution with 8.00% chitin, dissolved in the above mentioned mixture of trichloroacetic acid and formic acid.

A set of equal quantities of solution were prepared by dispersing chitin powder in the solvent mixture. These mixtures were subsequently subjected to ultrasonic treatment at room temperature, varying the ultrasonic power between 50W and 200W, and the treatment duration from 15 min to 2 hours. The ultrasonic treatment intervals and the resulted chitin molar mass are presented in Table 1.

The solution U4, made of 8% (w/v) chitin, prepared using an ultrasonic treatment of maximum power and duration, i.e. 200W and 2h, at room temperature, formed electrospun nanofibers with a very wide range of diameters (Figure 1), demonstrating an unstable electrospinning process due to the high viscosity and molecular mass of chitin.

The high viscosity and concentration of chitin caused a frequent breaking of the polymer jet and the fiber diameter resulted very nonuniform. This result suggests that increasing the power and duration of the ultrasonic treatment does not lead to an uniform chain fragmentation.

The solution remains viscous even after long treatment intervals, suggesting that the molecular chains are able to rapidly restore the interchain bonds broken by the ultrasonic energy.

In order to decrease the viscosity an increase the efficiency of the ultrasonic treatment, we decreased the chitin content of the blend.

Table 1. Ultrasonic treatment parameters and resulting molar mass of a 8% (w/v) chitin solution in 1:4 trichloroacetic acid and formic acid

Sample	Time (min)	M_w (kg/mol)
U1	15	1100
U2	30	1020
U3	60	970
U4	120	940

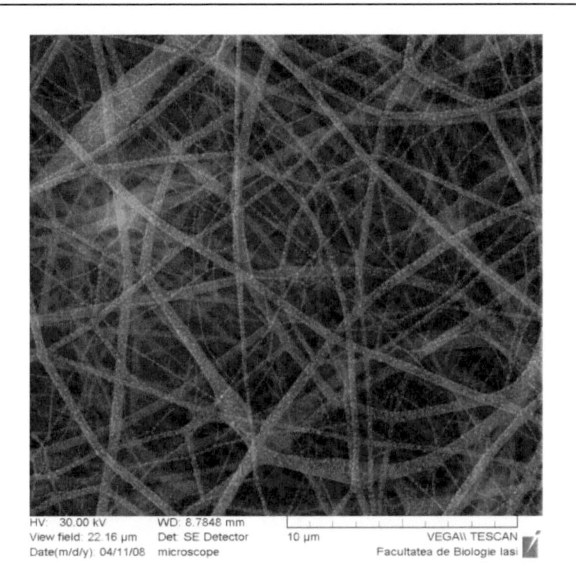

Figure 1. SEM image of nanofibers prepared by electrospinning a solution of 8.00% (w/v) chitin in 1:4 (v/v) trichloroacetic acid and formic acid.

Decreasing the chitin concentration to 1.33% (w/v), the 2h ultrasonic treatment at 200W became more effective (M_W=850kg/mol) and the electrospinning of the resulting solution lead to a slightly more stable electrospinning process, with a narrower distribution of fiber diameters. However, because the molecular length remained high and the molecules concentration low, the solvent accumulate sometimes under the action of surface tension and formed the beads seen of Figure 2. A higher quantity of solvent was more difficult to evaporate, and the fibers in Figure 2. manifested the tendency to bind at the connection points.

3.2. Chitin Depolymerisation by Microwave Treatment

The conclusion of the above described treatment is that supplying by ultrasound a supplementary energy to the solvent molecule aiming to make them more effective in breaking the hydrogen and the β-glycosidic bonds for cutting the chitin chains in shorter pieces, is not working. The ultrasonic energy involves only mechanical aspects of molecular movement. Ultrasound energy increases the brownian movement energy of solvent molecules, but this energy is dissipated homogenously on the chitin chains length.

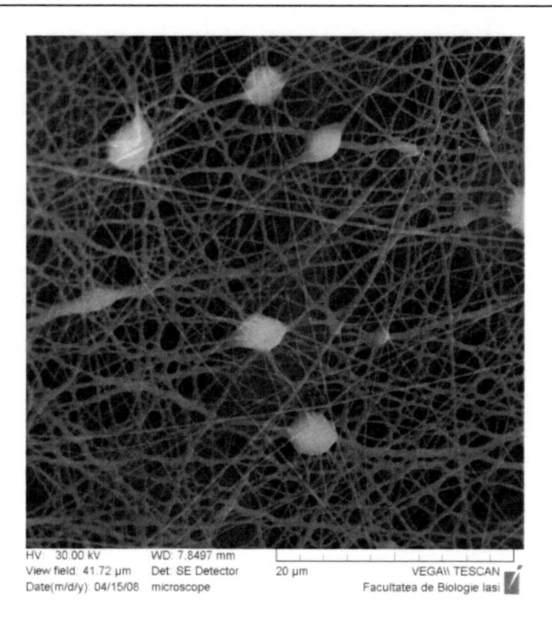

Figure 2. SEM image of nanofibers prepared by electrospinning a solution of 1.33% (w/v) chitin in 1:4 (v/v) trichloroacetic acid and formic acid.

In order to obtain an efficient molecular chain cutting, the supplementary energy should be focused on the bonds of interest, making them more susceptible for breaking. Microwave irradiation appears to be a more efficient form of stimulation, because it acts by stimulation the vibration of the bonds and enhances the electrical attraction between these activated bonds and the charged groups of the solvent molecules.

A second set of experiments were bound to finding the microwave treatment optimum parameters. Two sets of mixtures chitin powder/solvent were subjected to microwave treatments.

The CHEM STAR-2 microwave reactor has two possible modes of working. The first mode, named *Power Control* (PC), means that a percent of the full power (800 W) acts on the irradiated sample, while the sample temperature evoluates freely. In the second mode, called *Temperature Control* (TC), the reactor acts with pulses of full power, keeping the sample temperature at a constant value. In our experiments, a separate set of test have been performed for each working mode of the STAR-2microwave reactor.

In the first set of MW irradiation experiments, the reactor was working in the PC mode, following the parameters presented in Table 2. The investigated samples were appropriate quantities of chitin powder/solvent mixture, placed in standard Pyrex glass 10 ml tubes.

Table 2. Microwave treatment parameters, Power Control mode (PC), duration 3 minutes

Sample	Power (%)	T_{max} (°C)	M_w (kg/mol)
PC1	15	85	22.06
PC2	10	80	22.79
PC3	6	76	19.56
PC4	5	70	25.85

Table 3. Microwave treatment parameters, Temperature Control mode, T_{max} = 70°C

Sample	Power (%)	Time (min)	M_w (kg/mol)
TC1	100	15	19,56
TC2	100	5	26,03
TC3	100	3	29,46
TC4	100	1	36,03

The structural modifications of the chitin solutions after each treatment were investigated by FTIR spectroscopy. The recorded spectra of the MW treated chitin solutions in PC mode are shown in Figure 3.

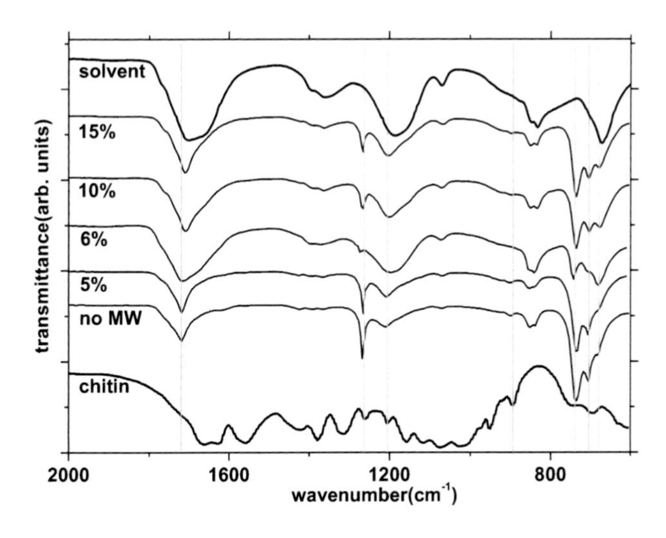

Figure 3. FTIR spectra of chitin dissolved in a mixture of trichloroacetic acid and formic acid, subjected to microwave treatment in power control mode.

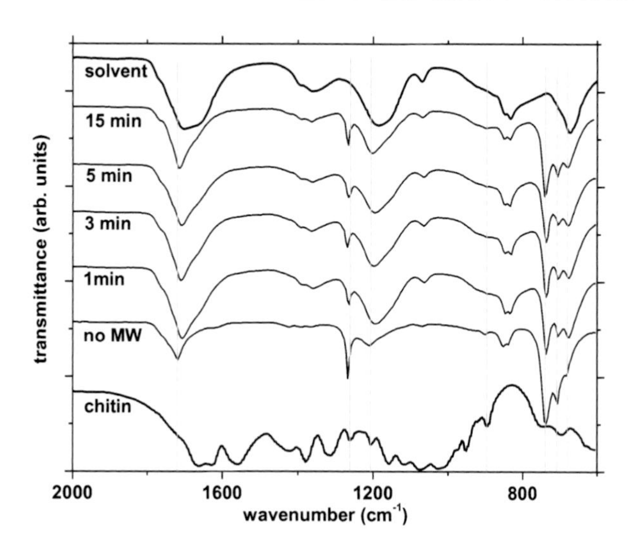

Figure 4. FTIR spectra of solution of chitin dissolved in trichloroacetic acid and formic acid, subjected to microwave treatment in Temperature Control mode, temperature T_{max}=70°C.

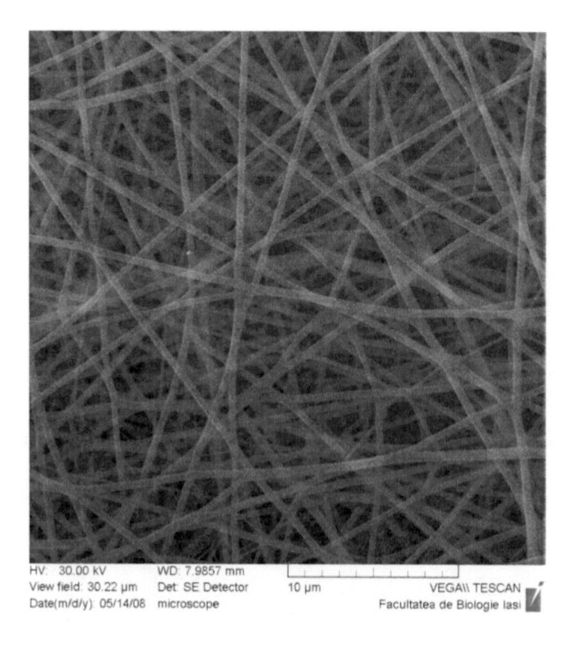

Figure 5. SEM image of nanofibers prepared by electrospinning a solution of 1.33% (w/v) chitin in 1:4 (v/v) trichloroacetic acid and formic acid, treated by microwave irradiation in Power Control mode, 6% of maximum power (800W), for 3 minutes.

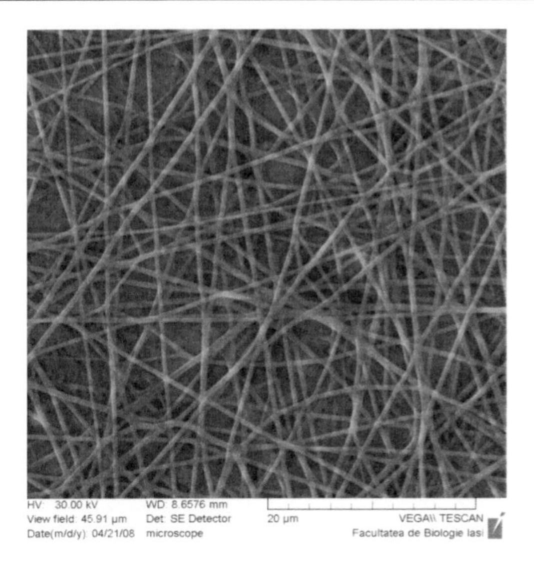

Figure 5. SEM image of nanofibers prepared by electrospinning a solution of 1.33% (w/v) chitin in 1:4 (v/v) trichloroacetic acid and formic acid, treated 5 min by microwave irradiation in Temperature Control mode.

A second set of MW irradiation experiments were performed on chitin/solvent mixtures irradiated with the reactor working in TC mode, keeping the temperature at 70°C and changing the treatment interval. The irradiation parameters are given in Table 3, and the FTIR spectra of the corresponding treated chitin solutions are shown in Figure 4. Samples of chitin solutions form PC-series and TC-series were mixed with 27% solution of gelatin dissolved in formic acid and tested for electrospinning. The electrospinning results showed that all the gelatin/chitin solution prepared with PC or TC microwave irradiated samples allowed stable electrospinning processes, but the nanofibers morphology presented some important differences depending on the microwave treatment type applied. Molecular mass measurements and FTIR spectra of the microwave treated chitin solutions, together with SEM images of the nanofibers electrospun from gelatin/chitin blends prepared using these chitin solutions, proved that chitin subjected to microwave irradiation undergoes both physical and chemical transformations. The physical transformations involve breaking of hydrogen bonds between macromolecular chains, increasing the distance between them and creating the possibility of breaking a chain into shorter pieces. The chemical transformations could consist of breaking etheric (positions 1-4 of the pyranosic ring) and estheric (amine group) bonds.

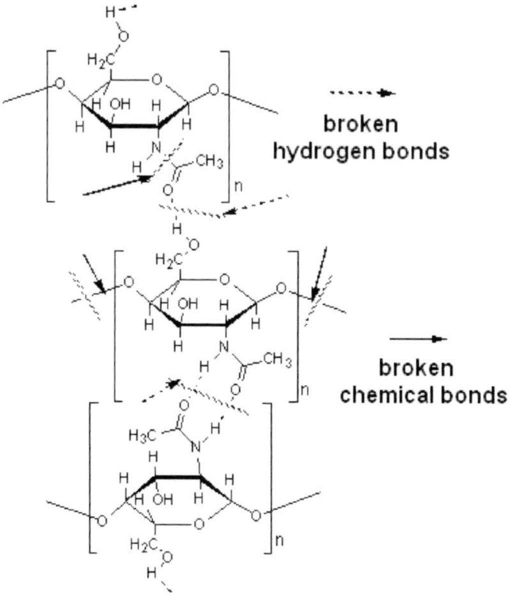

broken hydrogen bonds

broken chemical bonds

The shape of solution temperature time dependence in case of TC mode treatments, show that after a first interval of about 30 seconds, the chitin solution begins to absorb very fast the microwave energy, and after other 30 seconds the degradation of chitin ends, the temperature reaching a flat level.

The molecular mass of as received crab shell chitin, 1150±30 kg/mol, after MW treatment in solvent drops to 22±4 kg/mol, as proved by viscosity measurements. This result suggests that the MW treatment broke a macromolecular chitin chain in about 50 shorter pieces of close lengths. In this way, the resulting chitin keeps its physical and chemical macroscopic properties, but the solution properties change dramatically as a result of the new chain packing mechanisms possible between the shorter chains.

The as received chitin has long chains which tend to align parallel, giving rise to a long range order. In MW treated chitin solutions, the shorter chains are free to pack in a more diverse and flexible ways, as shown qualitatively in Figure 6.

The macromolecular packing mechanism presented in Figure 6.b. is supported by the FTIR spectra of the MW treated chitin solutions, which show changes in absorption bands of –NH, =C=O, -CH and CH_2 groups, as will be detailed in the following.

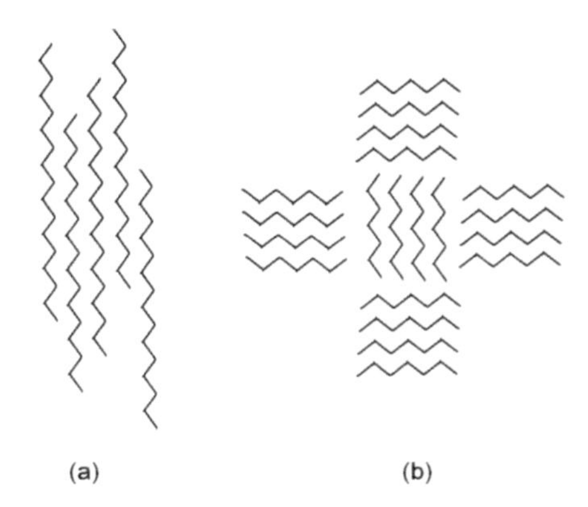

Figure 6. Macromolecular chains packing in crab shell chitin solutions: (a) as received (long chains), and (b) after microwave treatment (short chains).

The changes of hydrogen bonds and hydrogen-bonded groups can be observed on FTIR spectra [14]. In the FTIR spectra shown in Figures. 3 and 4, the peaks were assigned as follows: 1262 cm^{-1} to amide III (δ_{NH}), 705 and 680 cm^{-1} to amide V (γ_{NH}), 895 cm^{-1} to γ_{CH} and 740 cm^{-1} to ρ_{CH2} [16]. The deformation bands of the –CH group (γ_{CH}) show a decreased intensity after the microwave treatment of the chitin solution. This behaviour could appear because the anomeric carbon C_1 has a C_1-H oxidic bond, i.e. a β_{1-4} bound with the next ring, possible to appear only in a packing system as shown in Figure 6.b. The absorption band at 1750-1650 cm^{-1}, assigned to C=O stretching (amide I) of the MW treated chitin solutions appears broader as compared to the similar bands of the as received chitin, suggesting that a larger number of shorter chains interact by hydrogen and van der Waals bonds. Moreover, the C=O band of the as received chitin is centered at aprox. 1720 cm^{-1}, while the similar band of all MW treated chitin solutions is shifted with 10-40 cm^{-1} towards lower frequencies, suggesting an increased number of hydrogen bonds appeared between shorter chains (Figure 6.b). The NH amide group has absorption bands due to various types of deformations [17]. In case of MW treated chitin solutions, the intensity of NH bands at 1260 cm^{-1} and 705 cm^{-1} slightly decreases and shifts to lower frequencies, suggesting a (small) partial deacetylation of the NH amide group. Another chemical transformation that could lead to the observed changes in the NH band is the interaction of the NH group with an electrophile group of a chain fragment like CH_3-COOH. Low molecular fragments could explain the presence of some particles observed on the SEM images of the gelatin/chitin nanofibers prepared with non-optimal MW treated chitin solutions (Figure 7).

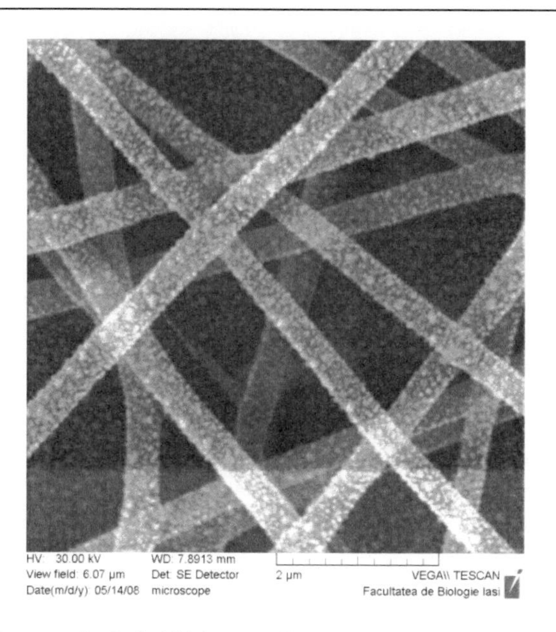

Figure 7. SEM image of gelatin/chitin nanofibers prepared by electrospinning a blend of 1.33% (w/v) chitin in 27% (w/v) gelatin/formic acid. The chitin solution was irradiated with microwave in Power Control mode, 15% of the total power (800W) for 3 minutes. The excess power lead to the formation of low molecular fragments accumulated on fibers surface during solvent evaporation.

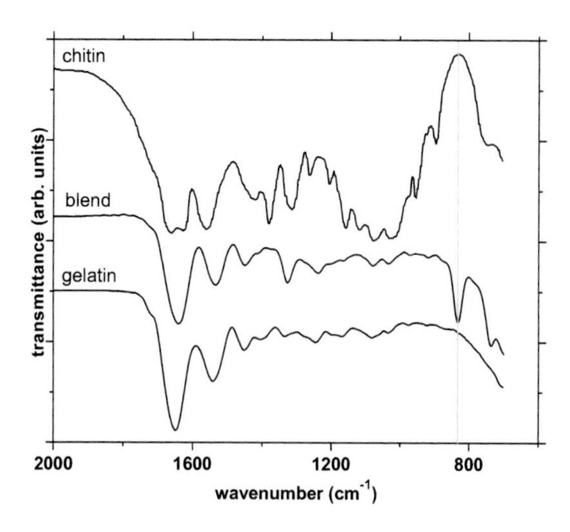

Figure 8. FTIR-ATR spectra of gelatin, chitin and gelatin/chitin nanofibers prepared using a chitin solution irradiated with MW in the TC mode for 5 minutes. The peak at 840 cm^{-1} suggests a chemical interaction between gelatin and chitin.

In Figure 8 is presented a FTIR-ATR spectrum of a gelatin/chitin nanofibers membrane prepared in the optimal conditions found in the present study, i.e. the chitin solution MW irradiated in TC mode for 5 minutes before mixing with the 27% gelatin/formic acid solution. As one can see in Figure 8, the absorption peak at 840 cm^{-1} does not belong to gelatin or chitin in their pure forms. This peak suggests a possible chemical bond between gelatin and chitin chains in the nanofibers structure. The mechanism responsible for this absorption is currently under investigation. The other bands of the gelatin/chitin nanofibers are basically the bands of gelatin, some of them having increased intensities as a result of interchain interactions with chitin.

CONCLUSIONS

In the present work, gelatin/chitin nanofibers with 1.33% (w/w) chitin have successfully prepared by electrospinning. In order to obtain an electrospinnable solution, the molecular mass of the as received crab shell chitin was decreased (chitin depolymerisation). Two possible depolymerisation methods have been tested, ultrasound and microwave irradiation. It was found that microwave irradiation gives the best results, with the reactor working in temperature control mode. The optimum solution was prepared by irradiating 10 ml of gelatin/solvent mixture, using pulses of 800W microwave power for 5 minutes, at 70°C. The gelatin/chitin blend prepared using this chitin solution gave the best quality of the electrospun nanofibers.

All the other gelatin/chitin blends prepared with microwave irradiated chitin solutions were electrospinnable. The main difference from the optimum nanofibers was the presence of low molecular fragments accumulated as a "dust" on the nanofibers surface. The low molecular fragments appeared as a consequence of excessive microwave power or irradiation time.

REFERENCES

[1] T. Nishido, K. Yasumoto, T. Otori, J. Desaki, *Invest. Ophth. Vis. Sci.* 29, 1887 (1988).

[2] C. S. Chen, M. Mrksich, S. Huang, G. M. Whitesides, D. E. Ingber, *Science* 276, 1425 (1997)

[3] N. Patel N, R. Padera, G. H. W. Sanders, S. M. Cannizzaro, M. C. Davies, R. Langer, *FASEB J.* 12, 1447 (1998).

[4] T. G. van Kooten, J. F. Whitesides, A. F. von Recum, *J. Biomed. Mater. Res.* 43, 1 (1998).

[5] H. K. Noh, S. W. Lee, J. M. Kim, J. E. Oh, K. H. Kim, C. P. Chung, S. C. Choi, W. H. Park, B. M. Min, *Biomaterials* 27 (2006) 3934–3944.

[6] S. M. Hudson and D. W. Jenkins, Chitin and Chitosan, in Encyclopedia of Polymer Science and Technology, John Wiley and Sons (2002).

[7] R. A. A. Muzzarelli, *Cell. Mol. Life Sci.* 53, 131 (1997).

[8] ISO 3219:1993. Plastics – Polymers/resins in the liquid state or as emulsions or dispersions –Determination of viscosity using rotational viscometer with defined shear rate.

[9] C. Tanford, Transport Processes – Viscosity, in Physical Chemistry of Macromolecules, Ed. C. Tanford, *John Wiley and Sons*, 391 (1961).

[10] M. U. Beer, P. J. Wood and J. Weisz, *Carbohydrate Polymers* 39(4), 377 (1999).

[11] A. Einbu, Characterisation of Chitin and a Study of its Acid-Catalysed Hydrolysis, PhD Thesis, Norwegian University of Science and Technology, Trondheim, April 2007.

[12] S. Ramakrishna, K. Fujihara, W. E. Teo, T. C. Lim and Z. Ma, An Introduction to Electrospinning and Nanofibers, World Scientific Publishing, (2005).

[13] G. A. F. Roberts, Chitin Chemistry, MacMillan Press (1992).

[14] E. Atkins, *Journal of Biosciences* 8(1-2), 375 (2007).

[15] J. Dong, Y. Ozaki, *Macromolecules* 30, 286 (1997).

[16] J. Brugnerotto, J. Lizardi, F.M. Goycoolea, W. Arguelles-Monal, J. Desbrieres, M. Rinaudo, *Polymer* 42, 3569 (2001).

[17] G. Cardenas, G. Cabrera, E. Taboada, S. P. Miranda, *J. Appl. Polym. Sci.* 93, 1876 (2004).

In: Electrospinning Process and Nanofiber... ISBN 978-1-61209-330-7
Editors: A.K. Haghi and G.E. Zaikov

Chapter 5

NANOTECHNOLOGY: A GLOBAL CHALLENGE IN HEALTHCARE

J. Venugopal[1] *, Molamma P. Prabhakaran[1], Zhang Y.Z[1], G. Deepika[1], V.R. Giri Dev[1], Sharon Low[2], Aw Tar Choon[2] and S. Ramakrishna[1]

[1]Nanoscience and Nanotechnology Initiative, Division of Bioengineering, National University of Singapore, Singapore
[2]StemLife Sdn BhD, Kuala Lumpur, Malaysia

ABSTRACT

Nanotechnology is the creation and utilization of materials, devices and systems through the control of substance in a nanometer scale. Nanobiotechnology creates a better understanding of cell biology because molecules in the cells are organized in nanoscale dimensions and they function as nanomachines. Nanomedicine is the process of diagnosing, treating and preventing diseases and traumatic injuries, relieving pain and improving human health. The high cost, together with a limited space for significant economies in the mass scale production of tissue engineered products has hindered widespread clinical application. In addition,

* National University of Singapore, Nanoscience and Nanotechnology Initiative, Division of Bioengineering, Block E3, #05-12 NUSNNI, 2 Engineering Drive 3, Singapore 117576, Phone : 65 – 6516 4272, Fax: 65 – 6773 0339, E-mail : engjrv@nus.edu.sg

presently available tissue engineered products still share some of the concepts of substitution medicine, where a laboratory grown 'spare part' is implanted in the body to compensate for lost tissue. Several of these recent developments in electrospun nanofibers are already at advanced phases of commercialization or clinical trials. By building pioneering achievements in tissue engineering, advanced therapies in the regeneration of pathological tissues to treat, modify and prevent disabling chronic disorders such as diabetes, osteoarthritis, diseases of cardiovascular and central nervous system are achievable. The vision for nano-assisted regenerative medicine ideally involves the development of cost-effective disease modifying therapies that will allow for *in situ* tissue regeneration. This article summarises the recent developments in electrospun nanofibers for healthcare applications.

Keywords: Electrospinning, polymers, nanofibers, stem cells, tissue engineering.

1. Introduction

Nanotechnology refers to one of the rapidly growing scientific disciplines studying and developing objects and materials with characteristic dimensions to resolve many of the disease related or organ damages in tissue engineering. Nanofibrous scaffolds are ideal for this purpose because their dimensions are similar to components in the extracellular matrix (ECM) and mimic its fibrillar structure, providing essential cues for cellular organization, survival function. Moreover, the elementary biological units in human body such as proteins, DNA or cell membranes are of nanometer scale. Therefore the application of nanotechnology in life science research, particularly at the cell level sets the stage for an exciting role in healthcare. Fabrication of nanofibers is one of the most important promising techniques for designing polymer nanofibers in tissue engineering. There are several scaffold fabrication techniques namely, electrospinning (random, aligned, core shell and vertical nanofibers), self assembly, phase separation, melt-blown and template synthesis [1]. Of these techniques, electrospinning is the most widely used technique and it also seems to be demonstrating promising results for tissue engineering applications. The role of biomaterials in tissue engineering is to act as a scaffold for cells to attach and organize into tissues. The ECM is a complex arrangement of proteins and polysaccharides such as collagen, hyaluronic acid, proteoglycans, glycosaminoglycans and elastin. These ECM components are constantly synthesized, secreted, oriented and modified by the cellular components that they support. Generally, the function of native ECM was only believed to be as a structural framework for tissues. The ideal tissue engineering scaffolds shall therefore mimic the ECM, which is the natural abode of cells. The structure and morphology of nonwoven nanofiber matrix was found closely match the structure of ECM of natural tissue [2, 3]. There are many naturally occurring nanoporous structure present in the human body further adds to the impetus for research and development in this area. Combined utilization of nanofibrous scaffolds together with drug delivery and gene therapy has great potential to provide improved tissue replacements [4].

Three dimensional biodegradable scaffolds, either natural or synthetic have been designed by electrospinning process serve as an excellent frame work for cell adhesion, proliferation and differentiation. Nanofibrous scaffolds not only serve as carriers for the delivery of drugs but also used as a scaffolds for engineering skin, bone, cartilage, vascular and neural tissue engineering. Collagen nanofibers for example, provide the cells with appropriate biological environment for cell growth and wound repair. Various polymeric nanofibers have been investigated as a novel wound dressing material and as haemostatic devices [5].

The high surface area of nanofiber matrix allows oxygen permeability and prevents fluid accumulation at the wound site. On the other hand, small pore size of nanofibrous matrix efficiently prevents bacterial penetration making them ideal candidates for wound dressings. The main goal in tissue engineering is to enable the body to heal itself by introduction of electrospun nanofibrous scaffolds, such that the body recognizes as its own and in turn uses it to regenerate "neo-native" functional tissues.

Moreover, the material when implemented *in vivo* should be removed via degradation and absorption, leaving the native tissue. Flexibility of electrospinning process also allows co-spinning of polymers with drugs or proteins thereby obtaining a nonwoven nanofiber mat that can act as a drug delivery matrix for an enhanced wound healing process [6]. Recently, electrospun nanofibers are being explored as scaffolds for nanodevices, biosensors, drug delivery and tissue engineering applications.

Electrospinning offers a simple set up with conduciveness to scale-up, providing diversity and control over the process, creating nanofibrous ECM analogues. Efficient organization and modification of nanofibers can be achieved by varying the electrospinning parameters. The geometry of grounded collector shall however determine the size and shape of the electrospun nanofibers. The large surface area to volume ratio of nanofibers combined with its porous structure favors cell adhesion, proliferation [7], migration [8] and differentiation [9], which are the desired properties for engineering tissues [10].

High porosity of nanofiber scaffolds provides more structural space for cell accommodation and facilitates efficient exchange of nutrient and metabolic waste between scaffold and the environment [6, 11]. Another important factor is the delivery of growth factors, which should be released in a sustained manner without loss of their bioactivity [12].

Mixing drugs with carrier polymers for electrospinning is a common approach for electrospinning of drug-incorporated nanofibers and also encapsulating bioactive molecules inside the nanofibers might be more beneficial. This manuscript summarizes the fabrication of electrospun nanofibrous scaffolds for several approaches for tissue engineering in healthcare applications.

2. Electrospinning

Electrospinning has been recognized as an efficient and well established technique capable of producing nanofibers by electrically charging a suspended droplet of polymer melt or solution [13].

Various polymers including synthetic ones such as poly(ε-caprolactone) (PCL), poly(lactic acid) (PLA), poly(glycolic acid) (PGA), poly(lactic-co-glycolic acid) (PLGA), polystyrene, polyurethane (PU), polyethyelene terephthalate (PET), poly(L-lactic acid)-co-poly(ε-caprolactone) (PLLA-CL) and biological materials such as collagen, gelatin, chitosan have been successfully electrospun to obtain fibers with diameters ranging from 3 nm to 5 μm [1, 11, 13].

The different parameters is to control the electrospinning process include the solution properties, controlled variables and ambient temperature. Utilizing a simple and inexpensive setup, this technique not only provides an opportunity for control over the thickness and composition of nanofibers but also controls fiber diameter and porosity of the electrospun nanofiber meshes. Typically nanofibers are collected as random, while aligned nanofibers with improved mechanical stability and degradation properties are also produced for specific applications [14].

Deposition of nanofibers on a static plate produces randomly oriented nanofibrous (100-650 nm) scaffolds; whereas aligned nanofiber (250-650 nm) mats are fabricated using rotating cylinder or disk collector with sharp edge as shown in figure (Figure 1.a,b).

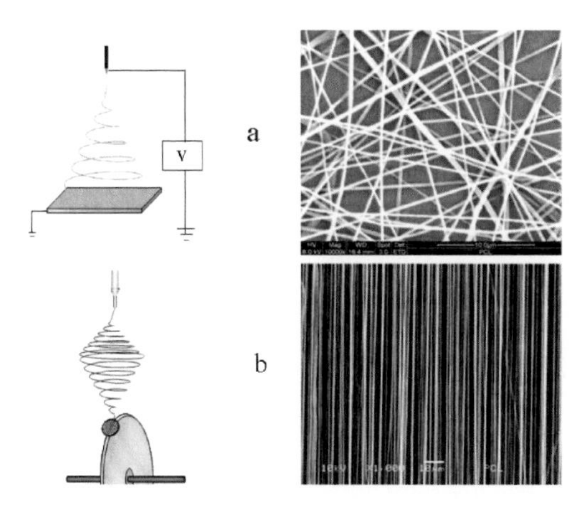

Figure 1. Schematics of electrospinning, (a) Random nanofibers produced by static collector, (b) Aligned nanofibers produced on disk collector in rotating wheel.

Co-axial electrospinning is a modification or extension of traditional electrospinning technique with a major difference being a compound spinneret used. Using the spinneret, two components are fed through different co-axial capillary channels and are integrated into core-shell structured composite fibers to fulfill different application purposes [15, 16]. For example; bioactive composite scaffolds are fabricated using collagen (imparting bioactivity) as the shell and PCL (synthetic polymer) as the core (Figure 2). Core-shell structured nanofibers (360-400 nm) prepared by co-axial electrospinning, have the advantages of being able to control the shell thickness and manipulate overall mechanical strength and degradation properties of resulting composite nanofibers, without changing its biocompatibility. Alternatively, core-shell structured composite nanofibers are functionalized for potential use in drug or growth factor encapsulation and release, and development of highly sensitive sensors and tissue engineering applications.

Water soluble bioactive agents are also incorporated into polymer meshes via co-axial electrospinning and their controlled release is reported (12). Moreover, growth factors and DNA could be readily integrated into the nanofibrous scaffolds for gene therapy and various tissue engineering applications [15, 16]. Other advantages of this technique include mild preparation conditions, high drug loading capacity and steady drug release properties [17]. Recent approaches include the development of multichannel microtubes, as an extension to two-channel co-axial electrospinning approach, providing route for fabrication of multifunctional nanofiber structures as well [18].

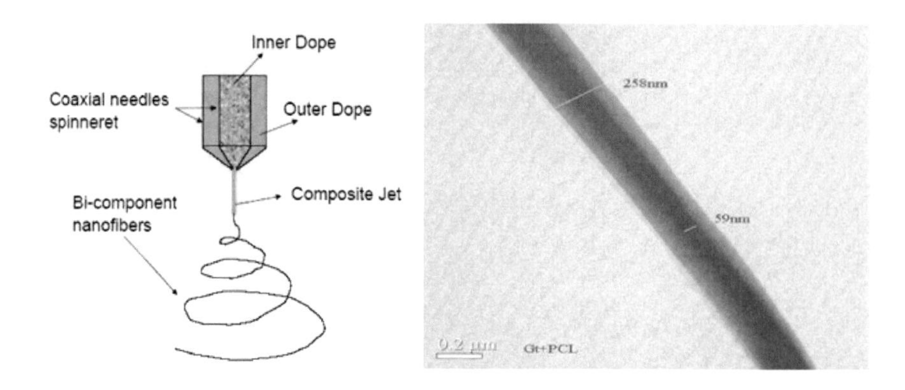

Figure 2. Co-axial electrospinning model for producing core-shell nanofibers.

3. Nanofibers in Tissue Engineering

Tissue engineering is the application of knowledge and expertise from a multidisciplinary field, to develop and manufacture therapeutic products that utilize the combination of matrix scaffolds with viable human cell systems or cell responsive biomolecules derived from such cells, for the repair, restoration or regeneration of cells or tissue damaged by injury, disease or congenital defects. Tissue engineering (TE) involve scaffolds or matrices to provide support for cells in order to express new extracellular matrix. The biocompatibility of scaffold materials actively participates in the signaling process for the requirement of safe degradation and also provides a substratum for cell migration into the defect sites of the tissue. TE is one of the most promising techniques for regeneration of tissues by autologous, allogeneic cells and tissue transplantation. However, autografts are associated with limitations such as donor site morbidity and limited availability. An alternative to autografts is allograft, which has potential to cause an immune response and also carry the risk of disease transfer [19]. Potential applications of TE are envisioned in the following fields of skin, cartilage, bone, blood vessel, cardiovascular diseases, nerve and soft tissues. Inert implantable or extracorporeal medical devices can rarely replace structure and function of natural tissues and organs [20].

The fibrillar structure of collagen is important for cell attachment, proliferation and differentiation function, and mimicking its structure may lead to engineered tissue more closely resembling native tissues [21]. Polymer nanofibers are important class of nanomaterials which are focused last ten years in the field of tissue engineering. Nanostructured materials are smaller in size falling around 1-100 nm range and has specific properties and functions related to the size of the materials. The development of nanofibers has enhanced the scope of fabricating scaffolds to mimic the architecture of natural human tissues at nanoscale. Engineering nanofibrous scaffolds mimicking the ECM as much as possible, both in terms of chemical composition and physical structure are achievable. A prominent feature of the diversified ECM structures is their nanoscaled dimensions and organization of different macromolecules. For example, the structural protein fibers with diameters ranging from several ten to hundred nanometers entangle forming a nonwoven mesh in a typical connective tissue, while the nano-scaled adhesive proteins like fibronectin and laminin provide specific binding for cell adhesion. Fibrous architecture of scaffolds with fiber diameters down to nanometer sizes are exactly suitable for replicating the physical structure of natural ECM. Such a structure meets the essential design criteria of an ideally engineering scaffold because they serve as a temporary ECM both architecturally and functionally, until the regeneration process occurs. Cells seeded on these nanofibrous scaffolds tend to spread, attaching multiple focal points and in some cases extend their filapodia along the length of nanofibers. In such cases, cells are seeded within the nanofibrous scaffolds, which may degrade or dissolve as the new tissue is formed for the repair of tissues [16]. Fabrication of nanofibers in multilayer structures may have potential for tissue engineering applications [22].

3.1. Wound Dressing and Skin Tissue Engineering

Skin is the body's largest organ, and functions foremost as a barrier, preventing pathogens from entering the body and also a sensory organ and a regulator for the water retention and heat loss. Several burns, pressure ulcers and other chronic ulcerations pose hurdles to caregivers and are costly to treat. Wound is described as a break or a defect of the skin from physical or thermal damage or due to underlying medical or physiological conditions [23]. Healing undergoes several stages: hemostasis, inflammation, proliferation and remodeling. The complex nature of wound healing requires the migration and proliferation of keratinocytes that are temporally regulated by numerous growth factors and their receptors that are upregulated in the wound environment. The complexity of wound environment has been recreated in human, bioengineered *in vitro* 3D tissues known as human skin equivalents, which have many morphologic and phenotypic properties of human skin. Dressings for human wounds aimed to protect, removal of exudates, inhibition of exogenous microorganism invasion and improved appearance. Normal bioactive wound dressing should fulfill several criteria for its optimal function: 1) biocompatibility in the absence of cytotoxicity, 2) easily applied to the wound surface, and easily removable after healing, 3) provides a moist environment and protects the wound against dehydration, 4) allows gas exchange between the wounded tissue and the external environment, 5) biofunctionalized to allow the slow release of bioactive agents. Tissue engineering of skin poses many challenges to researchers. Autografts and allografts are effective but their use is limited due to their inherent limited supply and cost. Current wound dressing materials from a variety of natural and synthetic sources. Many successful skin grafts utilize natural ECM proteins and collagen is cell adhesive protein extensively used in many successful wound dressing applications. Integra^{TM} is a temporary dressing composed of type I collagen and glycosamminoglycans (chondroitin 6 sulfate). Dermagraft^{TM} and TransCyte^{TM} both use cultured neonatal foreskin fibroblasts on synthetic polyglactin and nylon fiber, respectively. Apligraft^{TM} and Orcel^{TM} are both skin equivalents made from bovine collagen.

Recent technology in electrospinning has great potential for contributing in the field of skin tissue engineering. This process produces scaffolds that are highly conducive to cell infiltration, attachment and activity while also providing structural support for regenerating tissues. Electrospinning allows for the control of pore size and polymer diameter and thus provides a greater degree of control over cell infiltration. This makes electrospinning relevant for wound healing, as these properties are highly desirable in the development of wound dressings. Novel cost effective electrospun nanofibrous scaffolds are established for wound dressing and allogeneic cultured dermal substitute through the cultivation of human dermal fibroblast for skin defects [24, 25]. Other major issues in skin tissue engineering such as the delivery of nutrients, waste removal, gaseous exchange, protein transport, vascularization and tissue regeneration are governed by porous structure of the scaffold. However, a combination of growth factors together with porous structure of the scaffolds might substantially improve the skin regeneration efficacy. This can be achieved by a simple incorporation of growth factors during the scaffold preparation, either electrospinning process or obtaining a controlled release of growth factors via co-axial electrospinning technique. Recent developments including the use of growth factors and stem cell therapy are effective wound management strategies. An ideal scaffold shall possess excellent biocompatibility, controllable biodegradability and suitable mechanical characteristics. Electrospun PCL nanofibrous scaffolds were fabricated in our lab for the treatment of partial or full thickness skin defects. These nanofibrous wound dressing, due to their porosity and inherent property might achieve controlled evaporative water loss, excellent oxygen permeability and promote fluid drainage ability (Figure 3.a).

Allogeneic skin substitute composed of both fibroblast and keratinocytes is ideal; however the manufacturing cost is very high. Pleasing into account the potency of using fibroblast and keratinocytes leads to high manufacturing cost, we are using nanofibrous scaffolds with fibroblasts only in making allogeneic dermal substitute (Figure 3.b) for healing full thickness wounds and diabetic ulcers. The advantage of using this simple electrospinning methods for fabricating bioactive nanofibrous membranes for patients who will not heal well (diabetic ulcers) with low technology treatment will improve the process of triaging candidates for bioengineered skin and will increase the popularity and cost effectiveness of this approach.

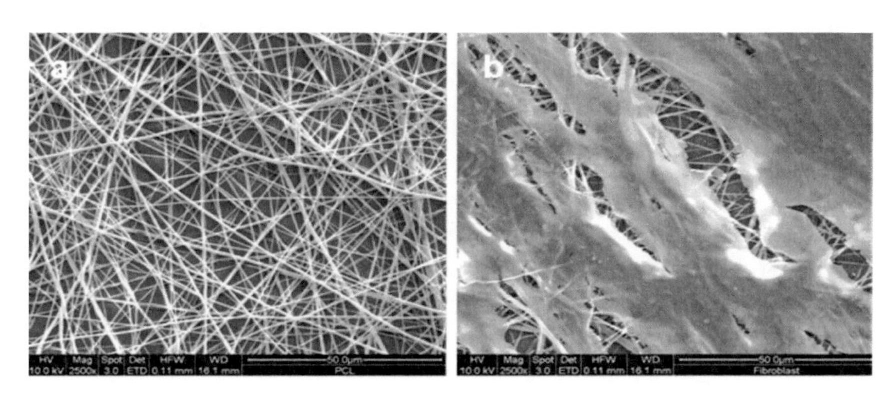

Figure 3. Nanofiber scaffolds for wound dressing and dermal substitute, (a) PCL nanofibers (350 nm), (b) Human dermal fibroblasts on PCL nanofibers (dermal substitute).

Addressing these issues can improve the nanofiber based product with the combination of desirable components (natural and synthetic polymers) into multifunctional scaffolds for pain relief, hemoglobin for oxygen delivery to healing tissues, and fibrinogen for hemostasis and adhesion and synthetic materials for additional mechanical stability, these functionalities have to increase in near future and these materials are expected to improve the quality of life and treatment efficacy [26].

3.2. Bone Tissue Engineering

Tissue engineering can be defined as the application of bioengineering disciplines to either preserve existing tissues or to allow tissues ingrowth by manipulating in vitro culture cells or tissues to activate the functions of defective or damaged body organs. Bone may be lost after trauma, cancer, fractures, periodontitis, osteoporosis, and infectious disease and presently very less techniques available for bone regeneration. Bone grafts are increasingly used, however they are plagued by high-failure rates of between 16% and 50% [27]. Autografts also present problems associate with a secondary surgery site as well as a limited supply and morbidity of the donor site [28]. The replacement of diseased bone tissues has taken a variety of forms: metals, ceramics, polymers and bone itself, none of which has proven ideal for tissue engineering. Biomaterials are permanent or biodegradable, naturally occurring or synthetic, need to be biocompatible, ideally osteoinductive, osteoconductive, integrative and mechanically compatible with native bone to fulfill their desired role in bone tissue engineering. These materials provide cell anchorage sites, mechanical stability, structural guidance, *in vivo* milieu and provide an interface to respond physiological and biological changes to remodel ECM in order to integrate surrounding native tissue [29].

Tissue engineered solutions containing cellular components and resorbable electrospun scaffolds offer enormous promise to restore tissue function without the need of the tissue removal. Recently, studies on three dimensional scaffold materials became a crucial element for bone tissue engineering. These scaffold materials were designed to mimic one or more bone forming components of autograft, in order to facilitate the growth of vasculature into material and provide an ideal environment for bone formation. The formation of bone can be roughly divided into three phases:

(1) proliferative phase, during which collagenous matrix is deposited,
(2) maturation phase, which is characterized by the activity of alkaline phosphate,
(3) mineralization phase, when the newly formed matrix begins to calcify.

These phases were influenced by collagen in the following manner: an increasing amount of collagen type III coating gave rise to an increase in proliferation and synthesis of collagen, both of which are characteristics for early phase of bone formation. A proper balance between osteogenesis and biomaterial resorption is however crucial for bone formation. A crucial point for a scaffold to be successful, especially in bone tissue engineering, is the combination of structural/mechanical properties of polymer structure and biological activities, all of them playing a critical role in cell seeding, proliferation and new tissue formation. Collagen provides an inherently good biocompatibility with cells and collagen based implants are well known for their feasibility in promoting tissue regeneration [30, 31]. Hydroxyapatite ($Ca_5(PO_4)_3OH$) considered as a structural template for bone mineral phase and also a major inorganic mineral component of bone and commonly used as bioceramic filler in polymer based bone substitute [32, 33] because of its high bioactivity and biocompatibility (Figure 4).

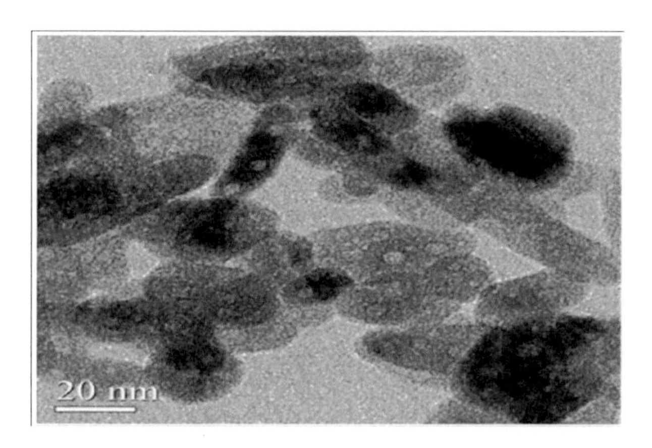

Figure 4. Transmission Electron Microscopic Image of nanohydroxyapatite (52 nm).

HA, the most stable calcium phosphate in natural environment is also having the inherent property to osteogenetically integrate into the bone [34]. It has a prominent affinity for regulating cell function and promoting osteogenesis and mineralization of bone. Nano to microscale alignment of nanohydroxyapatite/Collagen (nHA/Col) composite was similar to that of natural bone and thus might have been identified as 'bone' by the attached cells. The ECM is important, not only as a structural component for supporting cells, but also as a suitable microenvironment that influences cell-function. A number of short sequences in proteins located in the ECM have been recently been identified to play an important roles for bone regeneration, including mesenchymal stem cells (MSC) and osteoblast for proliferation, migration and differentiation.

Natural bone ECM composite consists of type I collagen and hydroxyapatite. The HA is orderly deposited within the nanofibrous collagen matrix and also initiate oseoconductivity and bone bonding ability [35]. Composite materials often show a good balance between toughness, strength and improved characteristics compared to individual components. However, the use of HA alone is limited due to brittleness and difficulty to process complex shapes for bone tissue engineering. PCL has been one of the most popular polymer used for bone tissue engineering scaffolds because of its biocompatibility, slow degradation and ease of electrospinning from a variety of solvents. Mesenchymal stem cells from rats cultured on electrospun PCL scaffolds, supplemented with osteogenic media for 4 weeks, showed complete penetration of the scaffolds with the formation of multilayers. Mineralization and type I collagen detected by von Kossa protocol and immunostaining, respectively [36]. Chitin is the planet's second most abundant carbohydrate; chitin and its deacetylated product chitosan have been long recognized for their antifungal properties and ability to induce protease inhibitors in plants [37]. Chitosan has long been considered as one of the most attractive natural biopolymer matrices for bone tissue engineering owing to its structural similarity to the glucosaminoglycan found in bone, biocompatibility, biodegradability and excellent mechanical properties [38]. A preliminary investigation into a 90:10 PEO:chitosan blend showed that the scaffold was non-toxic and osteoblast spreading in culture was optimal, and spreading filapodia for attaching to grew along the direction of the polymer nanofibers [39]. Zhang *et al.* fabricated the scaffolds in a two-step approach that combines an *in situ* co-precipitation synthesis route with electrospinning process to prepare a novel type of biomimetic nanocomposite nanofibers of hydroxyaptite/chitosan (HA/CTS). The electrospun composite nanofibers of HA/CTS, with compositional and structural features close to the natural mineralized nanofibril counterparts, are of potential interest for bone tissue engineering [39]. The results of HA/CTS indicate that notwithstanding the occurrences of an initial inhibition, the HA incorporated nanofibrous scaffolds as compared to CTS alone scaffolds appeared to have significantly stimulated the bone forming ability as shown by the cell proliferation, mineral deposition, and morphology observation, due to the excellent osteoconductivity of HA. In another study, electrospinning of PCL/nHA/Col and PLACL mixture produced highly porous nanofibrous structure with high surface area and provided sufficient mechanical strength for handling scaffolds. This fibrous architecture mimicked natural ECM and assisted in maintaining a normal phenotype and mineralization of osteoblast cells [40-42].

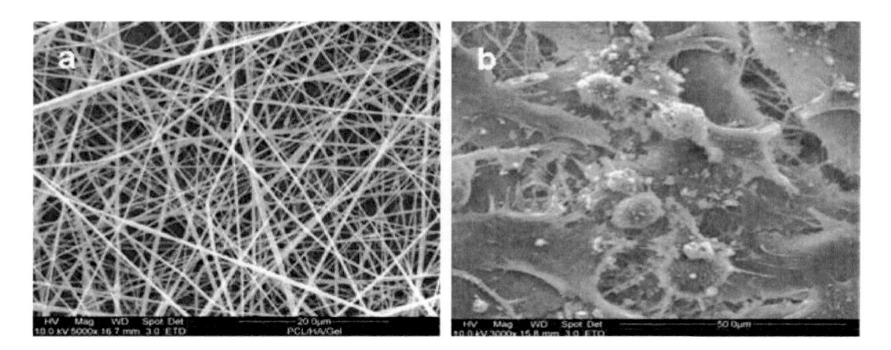

Figure 5. Nanofiber scaffolds for bone tissue engineering, (a) PCL/nHA/Gelatin biocomposite nanofibers (356 nm), (b) Human fetal osteoblasts with mineral deposition on biocomposite nanofiber scaffold.

The surface topography of nanostructured substrate plays a critical role in regulating initial cell behaviors, such as cell adhesion, which can also influence cellular viability and proliferation. Deposition of a biomimetic apatite layer throughout the porous structure of the 3D scaffolds in an effective method for controlling surface topography and chemistry within large, complex structures. Mechanical properties of the electrospun PCL/nHA/ Gelatin nanofibrous structure (Figure 5.a) was enhanced by PCL, while cell proliferation was supported by gelatin, and HA acted as a chelating agent for mineralization of osteoblast cells (Figure 5.b) for bone regeneration [43-45]. These biocomposite nanofibrous structures have great potential for bone filling and bone tissue regeneration.

3.3. Artificial Blood Vessel Engineering

The challenge for producing vascular grafts is to engineer vascular replacements that can withstand the high pressure and flow rate of the blood stream [46]. Cardiovascular diseases are the leading cause of mortality in western countries and are becoming the one in developing countries. Most ischemic diseases like atherosclerosis require a revascularization procedure used for vascular regeneration. Normally, they are either treated by percutaneous coronary intervention with or without stent replacement, or bypass surgery [47]. Electrospun scaffolds are economical, easy to produce and can be fabricated into highly interconnected porous scaffolds permits cellular infiltration and vascular ingrowth. Electrospinning offers the potential for greater control over composition, mechanical properties and structure of a graft, making it easier to match the compliance of the synthetic scaffold to that of the native artery. Recent approaches in blood vessel tissue engineering are to develop biologically and mechanically stable, immunologically safe and thrombosis resistant artificial blood vessel [48].

A new generation of biomaterials is being developed to mimic the structure and characteristic of native ECM, such as fibrillar structure, viscoelasticity, cell addition domains, growth factor binding and proteolytic sensitivity. Such materials are attractive because, in principle, their properties can be readily controlled while mimicking many of the critical biological functions of the native ECM, which are largely lacking from synthetic polymers such as PGA. The technique of electrospinning has been used to produce fibers with diameters on the order of those found in native ECM. Vascular tissues are subject to 4 principle hemodynamic forces: (1) shear stresses, tangential frictional forces acting on endothelial cells (ECs) as a result of blood flow and on smooth muscle cells (SMCs) as a result of transmural interstitial flow, (2) luminal pressure, a cyclic normal force attributable to blood pressure, (3) mechanical stretch, a cyclic circumferential stress caused by blood pressure and, (4) tension in the longitudinal direction. All of these forces have been shown to act both independently and synergistically to modulate the behavior of vascular tissues. It is the clinical urgency for improved blood vessel substitutes, especially for small-diameter applications that drives the field of vascular tissue engineering.

Poly(ethylene terephthalate) (DRACON) or poly(tetrafluoro ethylene) are preferred for large vessel reconstruction, they are unsuitable for small caliber grafts because the patency remains poor [49]. Studies suggested that SMCs, once implanted the scaffold developed the function, shape, morphology and cellular architecture of normal vessel. ECM surrounding the vascular cells complies and combines to provide biomechanical properties of the tissue. Electrospinning of P(LLA-CL) polymer fibers into a vascular conduit has been demonstrated to be a potential technique that not only creates scaffolds simulating the ECM, but also contributes tailorable mechanical properties (Figure 6). Topographically aligned submicron fibers have similar circumferential orientations to the cells and fibrils found in the medial layer of the native artery. Because of these similarities cell viability is often superior on electrospun scaffolds than on other substrates. P(LLA-CL) composed of aligned and random nanofibers were fabricated to mimic the native ECM and favorable interaction with SMCs and ECs between the nanofibrous scaffolds was demonstrated by He *et al.* [50,51].

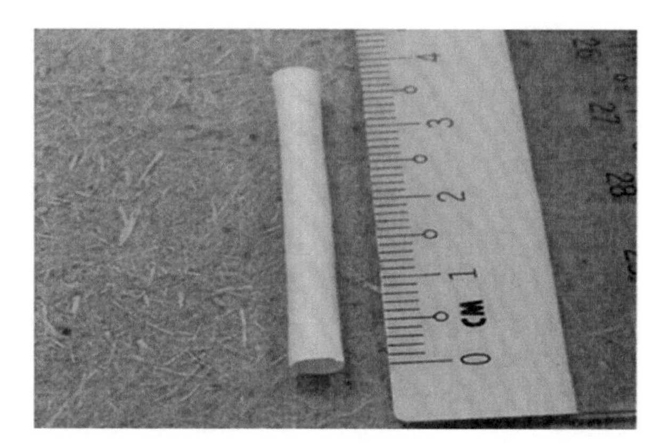

Figure 6. Artificial blood vessel fabricated with P(LLA-CL) copolymer nanofibers (Courtesy by Teo Wee Eong, NUSNNI, National University of Singapore).

Shum-Tim *et al.* [52] have seeded a polyglycolic acid-polyhydroxyalkanote (PGA-PHA) co-polymer directly with mixed population of ovine carotid ECs, SMCs and fibroblasts. The PGA inner layer was used to promote cell adhesion and tissue formation, whereas the PHA layer provided the necessary mechanical strength with a longer degradation period of 52-weeks. The tissue engineered grafts were used to replace infrarenal aortic segments in seven lambs and all remained patent for upto 5 months. The first clinical application of an artificial vessel based on a biodegradable scaffold was reported by Shin'oka *et al.* [53]. Cells isolated from a peripheral vein were seeded on PCL-PLA co-polymer tube reinforced with woven PGA. The grafts were used for the reconstruction of an occluded pulmonary artery in a 4 year old girl. Seven months after implantation, the patient was doing well, with no evidence of graft occlusion or aneurysm formation [54, 55]. Aligned nanofibrous scaffolds prepared from synthetic biodegradable polymers mimic natural ECM and create a defined architecture replicating an *in vivo* like vascular structure. In an attempt to produce aligned P(LLA-CL) co-polymer nanofibrous scaffolds (Figure 6.a) for artificial blood vessel engineering, our research group utilized a novel 'sharp edge rotating wheel' method successfully [56, 57]. Cell culture results of SMCs on these nanofibrous scaffolds indicated that the cells attached and migrated along the direction of aligned fibers and expressed spindle-like contractile phenotype organization and distribution of smooth cytoskeleton protein (α-actin) inside the cells (Figure 6.b).

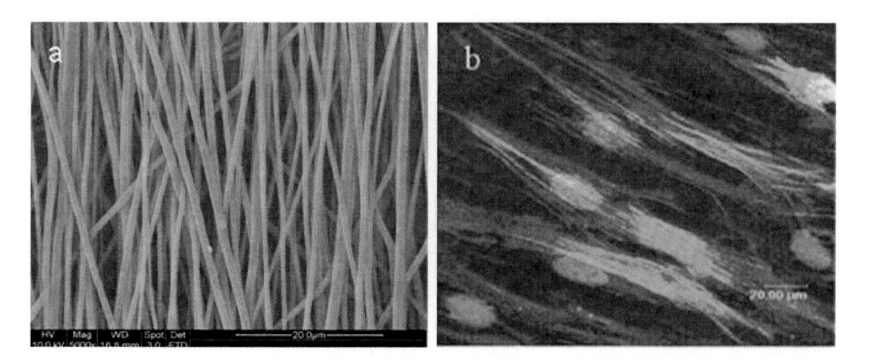

Figure 5. Nanofiber scaffolds for vascular tissue engineering, (a) Aligned P(LLA-CL) nanofibers (300 nm), (b) Expression of α-actin cytoskeleton protein in smooth muscle cells on aligned nanofibers.

The chemical composition, construction parameters and biomechanics of a vascular graft influence its interaction with its host. The differences of graft materials, porosity, compliance, electrical charge and surface texture all contribute to the magnitude and surface characteristics of the body's inevitable 'foreign body' reaction. SMCs and ECs were found to interact and integrate well with nanofibrous scaffolds forming a 3D cellular network [58]. This novel nanofibrous scaffold could help to understand vascular biology by providing a new tool for the study of ECM system in the formation of new blood vessels. Analyzing the balance point between support provided by the scaffold and the formation of tissue with sufficient mechanical integrity to support itself remains a big challenge in cardiovascular tissue engineering. These studies demonstrate that PCL/Collagen, P(LLA-CL) co-polymer nanofibrous scaffolds influence cell adhesion, migration, proliferation and support the formation of 3D blood vessels *in vitro* [7, 59]. One can, therefore envisage an advanced composite 3D scaffold that incorporates nanofibers and 2D surfaces for the growth and control of different cell types required for the successful development of vascular grafts by tissue engineering.

3.4. Neural Tissue Engineering

The nerve tissue engineering represents a significant challenge in the field of tissue engineering. Research in understanding and controlling nerve regeneration is one of the most urgently needed areas of modern medicine [26]. Several options like suturing of severed ends together, using autograft or allograft between the nerve ends or regeneration through a biological conduit are conducted for nerve repair. One day, there will be some breakthrough, and those with nerve injury (including such central nervous system injuries as those of the spinal cord) will have some hope at restoring normal function. For large nerve defects, autografts still remain the "gold standard" for nerve repair and this inevitably involves sacrificing at least one nerve. However, the limited availability of donor nerves and drawbacks of second operation for nerve harvesting lead to the idea of developing tissue-engineered nerve grafts as substitutions [60].

Traditionally the biomaterials based approach to bridging nerve gaps uses tubular constructs called as 'guidance channels". The purpose of introducing a nerve guide between the stumps of a transected nerve is to provide mechanical guidance with stimulating environment for the advancing axons [61].

Integra NeuroSciences (Plainsboro, NJ) developed a semipermeable collagen tube "NeuraGen" nerve guide and it has been clinically used for repairing peripheral nerve injuries [62]. Biodegradable materials are preferred over non-degradable artificial nerve guides since the non-degradable materials resulted in excessive scar formation, inflexibility and lack of stability.

In addition to scaffold composition and surface features, properties such as the degradation rate, permeability and porosity also play an important role in regulating nerve regeneration. Tailoring the flexibility of biodegradable polymeric materials was readily possible by altering their chemical and engineering properties. Such nerve guides allow neurotrophic communication between the nerve stumps and provides physical guidance to regenerative axons. Moreover, chemical stimulants, biological cues and physical guidance cues can further improve the efficacy of nerve guides.

A variety of biomaterials, in particular the polyesters, have been investigated for their suitability in nerve tissue engineering. Studying the *in vivo* regeneration using PLGA nanofibers, Bini *et al*. [63] implanted the nanofiber guides to the right sciatic nerve of rats and found no inflammatory response together with successful nerve regeneration. PLLA nanofibers were functionalized with laminin by Koh *et a*l. [64] via three different methods, and they found enhanced axonal extensions using electrospun PLLA-laminin nanofibers, showing them as suitable substrates for nerve tissue engineering. Various growth factors have also been added to the artificial nerve graft to influence regeneration and the controlled release of NGF, EGF and bFGF from polymer conduits was found to enhance the growth of axons [65-67].

Few reports are available on the contact guidance of NSCs cultured on electrospun porous scaffolds until recently. Yang *et al.* [68] fabricated random and aligned PLLA polymeric nano and microfibers for understanding the contact guidance of NSCs cultured on PLLA scaffolds. The NSC differentiation rate was found higher on nanofibers than on microfibers, independent of their fiber alignment. Aligned nanofibers highly supported NSC proliferation and enhanced the neurite outgrowth compared to microfibers and random nanofibers.

However, the cellular mechanism behind the directional effects of substratum contours on the behavior of NSCs need to be explored further. Further to this study, Ghasemi *et al.* [69] fabricated random and aligned PCL/gelatin (70:30) nanofibers by electrospinning and they found enhanced nerve cell proliferation and differentiation with improved neurite outgrowth on aligned PCL/gelatin nanofibers than PCL nanofibers.

Among the many factors that stimulate and control axonal regeneration, the most important influence derives from the local environment of the lesion [61]. Adequate substrates of trophic factors provided by reactive Schwann cells (SCs), macrophages and ECM within the degenerated nerve are required for axonal regeneration. The success of SCs in peripheral nerve regeneration (PNR) results partly due to their production of neurotrophic factors and cell adhesion molecules such as NGF and laminin, which mediate neurite attachment and growth [62].

In comparison to pure PCL nanofibers, PCL/collagen nanofibers were showed to improve SC migration and neurite orientation by Schnell *et al.* [70]. Few studies also reported the significant increase in the rate of peripheral nerve regeneration in guidance channels filled with laminin rich matrigel [71, 72]. Conduits incorporated with oriented nanofibers assist to recreate the bands of Bungner and these nanofibrous scaffolds were found to promote SC migration after peripheral nerve injury, with functional outcomes comparable to autografts.

Aligned PCL nanofibers were utilized by Chew *et al.* [73] to study their potential in providing contact guidance to human SCs. Cell cytoskeleton together with nuclei was found to align and elongate along the fiber axes, emulating the structure of Bungner bands. Among the natural polymers, chitosan a polysaccharide derivative of chitin, improved nerve regeneration by facilitating migration and proliferation of SCs [74].

In another study, biocomposite PCL/Chitosan nanofibers with fiber diameters of 190 nm were fabricated by electrospinning process (Figure 7.a). *In vitro* evaluation of these scaffolds using rat SCs showed an improved cell proliferation compared to electrospun PCL nanofibers (Figure 7.b), providing them as ideal substrates for peripheral nerve regeneration [75-77].

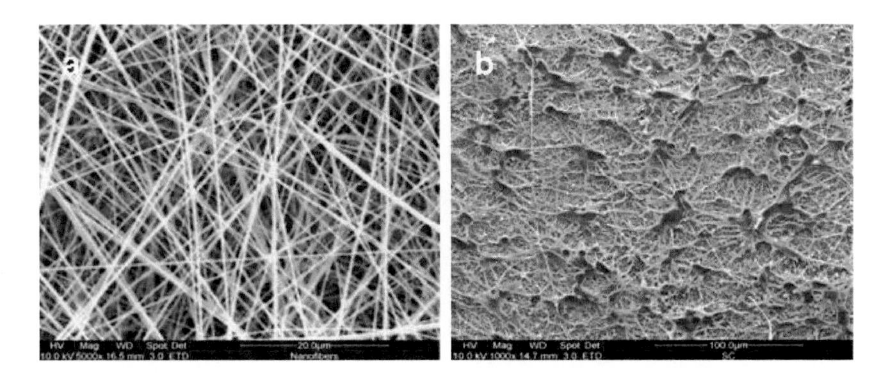

Figure 7. Nanofibrous scaffolds for neural tissue engineering, (a) SEM micrograph of PCL/Chitosan nanofibers (b) Schwann cells on biocomposite PCL/Chitosan nanofibrous scaffolds.

Guenard *et al.* [78] reported that SCs pre-seeded in nerve guidance channels showed enhanced peripheral nerve regeneration and over longer distances. SCs play a dual role in nerve regeneration, such that they serve as a physical framework for regenerating nerves (band of Bungner), providing ECM, adhesion molecules facilitating cell attachment [79] and also as a source of various trophic factors for regenerating axons [80]. The advantages of seeding Schwann cells within a biosynthetic collagen guide for the repair of severed nerves was also evaluated by Ansselin *et al.* [81], considering the use of autologous cells in human nerve repair. The concentration of SCs seeded into a conduit was found to influence the regenerative capacity of axons [81]. Collagen nerve guides seeded with more than 5×10^5 cells gained more functional recovery in mere 20 weeks compared to the nerve guide seeded with less than 5×10^5 cells.

Successful nerve regeneration can be achieved by tissue engineering scaffolds that provide mechanical support to the growing neuritis, while physical or chemical binding of ECM proteins on biodegradable electrospun scaffolds (surface modification) will further enhance their outgrowth of neural cells in nerve regeneration. Studies by other researchers showed neurite outgrowth along aligned SCs incorporated with materials such as fibronectin, collagen and micropatterned polymers also suggested the essential role of SC orientation and topography in guiding axons [82, 83]. The ECM, an environment desired to be mimicked is a complex mixture of polysaccharides and proteins, such that multi-component systems, if engineered to promote nerve regeneration might be of advantageous over single-component systems. ECM proteins are involved in specific interactions with neural tissues to produce biomimetic scaffolds suitable for nerve regeneration [84]. Collagen, laminin, hyaluronic acid and fibronectin are ECM components used for nerve reconstruction, while interactions between multiple ECM components might be important as well. Collagen filaments were found to guide the axonal regeneration of rat sciatic nerve but their disadvantages include the high cost and mechanical weakness of collagen. Microgrooved poly(D,L-lactic acid) substrates adsorbed with laminin or immobilization of fibronectin derived peptides might also provide cues for nerve guidance and neurite outgrowth [83]. In future, fabrication of electrospun natural and synthetic polymer blends with the controlled addition and release of growth factors would be a potential nanofibrous scaffolds for nerve tissue engineering.

4. Stem Cells and Nanofibers

Stem cells are the functional elements of tissue engineering and regenerative medicine, though the applications of living cell therapy present several challenges [85]. Stem cells have been isolated from adult and embryonic tissues.

An adult stem cell is defined as an undifferentiated cell found among differentiated cells in an organ or tissue, which can renew itself and can differentiate to yield major specialized cell types of organ or tissues. These adult stem cells maintain and repair the tissue in which they are found. They can be harvested from adipose tissue, bone marrow, mammary tissue, CNS, olfactory bulb etc. Adult stem cells also possess the ability to transdifferentiate, i.e, they switch their specific developmental lineage to another cell type of different lineages [86].

The molecular mechanism behind this transdifferentiation process is however not clearly known. The most commonly studied stem cells are the bone marrow, especially the mesenchymal and hematopoietic. Under controlled conditions, the mesenchymal stem cells (MSCs) can differentiate into multi-mesenchymal lineages such as the osteoblast, chondrocyte, cardiomyocytes and fibroblasts [87].

Possessing the multi-differentiation capability and expandability, MSCs are regarded as a highly promising cell source for tissue engineering applications. Recent studies suggested that the embryonic stem cell (ESC) transplantation as a means of treating peripheral nerve injuries and possibly spinal cord injuries [88, 89].

Functionalized electrospun nanofibers with growth factors could enhance the differentiation of ESC cells into neurons and oligodendrocytes [90]. Electrospun nanofibrous scaffolds were utilized by Xie *et al.* [91] for enhancing the differentiation of mouse ESC cells into neural lineages and further promoting and guiding the neurite outgrowth. Such strategy of a combination of electrospun scaffolds together with ESC derived neural progenitor cells might lead to better nerve repair.

Bone marrow mesenchymal stem cells (BM-MSCs) and umbilical cord blood cells are currently of fascination to medical world. The differentiation potential of MSCs into hepatocytes on PCL-collagen-polyethersulfone scaffolds was studied by Kazemnejad *et al.* [92]. The capability of hepatocyte cells to produce albumin, urea, serum glutamic, pyruvic, transaminase and serum oxaloacetate aminotransferase on the scaffolds further confirm the supporting role of the functionality of cells grown on nanofibrous scaffolds. On the other hand, osteoblastic differentiation potential of MSCs on PLLA and collagen nanofibers was studied by Schofer *et al.* [93]. These researchers identified the advantages and disadvantages of more stable PLLA fibers and cell supportive collagen fibers with respect to osteoblastic differentiation for bone tissue engineering. Inducing MSCs for cardiac differentiation *in vitro* is most commonly being carried out by exposure to DNA de-methylating agent 5-azacytidine [94, 95] and Balana *et al.* found a variety of cardiac specific genes and peptides being expressed [95]. *In vitro* manipulation of cells prior to transplantation such as the cell seeded scaffolds, preconditioned in culture and their genetic transfection are also carried out by Davis *et al.* [96]. The multi-lineage differentiation potential of MSCs on PCL nanofibrous scaffolds were also studied by Li *et al.* [97]. The following is a list of steps in successful cell-based treatments that scientists will have to learn to precisely control to bring such treatments to the clinic. To be useful for transplant purposes, stem cells must be reproducibly made to: i) Differentiate into the desired cell types, ii) Survive in the recipient after transplant, iii) Integrate into the surrounding tissue after transplant, iv) Function appropriately for the duration of the recipient's life, v) Avoid harming the recipient in any way. In order to avoid the problem of immune rejection, scientists are experimenting with different research strategies to generate tissues that will not be rejected.

Human umbilical cord blood (UCB) cells have become an alternative source of hematopoietic precursors for allogeneic stem cell transplantation in children with inborn errors or malignant diseases [98]. Hematopoietic stem cells (HSCs), the main stem cell population of the bone marrow, are used for the treatment of sickle cell anemia, thalassemia, aplastic anemia, leukemia, metabolic disorders and certain genetic immunodeficiencies [99]. The cord blood stem cells show a higher proliferative capacity and expansion potential. Allogeneic stem cell transplantation can be limited by the lack of suitable bone-marrow donors and the risk of graft-versus-host diseases. The percentage of stem cells is higher in cord blood than in bone marrow and mobilized blood but the absolute number is lower than in other stem cell sources. The advantages of UCB stem cells over the other sources are: (1) easy to recover, (2) no health risks for the patients, (3) immediate disposition at the cryobank, (4) low incidence of rejection of the transplant, (5) high cellular plasticity, (6) low possibilities of transmission of viral diseases, (7) low cost of the procedure and (8) easy possibilities to create cord blood banks to store samples. One major advantage of UCB stem cells in comparison with peripheral blood stem cells or bone marrow is the reduced incidence of acute graft-versus-host disease caused by cord blood graft. Stem cells live 'happily ever after' in various tissues in the body where they can contribute to repair processes physiologically [100]. Transplantation protocols for adults have been restricted by limited number of progenitors contained in one cord blood harvest and due to this motivation, expanding HSCs *ex vivo* to get sufficient number of cells for transplantation became essential and successful. Several studies have shown that the nanofibrous scaffolds can enhance cellular responses like cell adhesion and cell phenotype maintenance [1, 29].

Studies conducted on three dimensional PCL nanofibrous scaffolds proved them as potential substrates for support and maintenance of multilineage differentiation potential of MSCs (human bone marrow) *in vitro* [97]. Investigations on the nanotopographical cues and various chemical cues on the nanofiber surface that can synergistically influence HSC adhesion, proliferation and phenotypic maintenance were also carried out. The highest expansion efficiency of $CD34^+$, $CD45^+$ cells and colony forming unit potential was observed in functionalized electrospun nanofibrous scaffolds [101]. Studies by Chua *et al.* [102] utilized amino groups conjugated as spacers to nanofiber surfaces and the cell-substrate interaction was found to regulate the HSC/progenitor cell proliferation and self renewal in cytokine supplemented expansion. Aminated nanofibers and PCL/Collagen nanofiber mesh were found to enhance the HSC-substrate adhesion and expansion of forming progenitor cells (Figure 8). This study provides the basis for further investigation of the effect of more specific cell adhesion molecules, e.g.

fibronectin, in combination with the nanofiber substrate, on HSCs adhesion and expansion *ex vivo* to solve metabolic disorders of the patients.

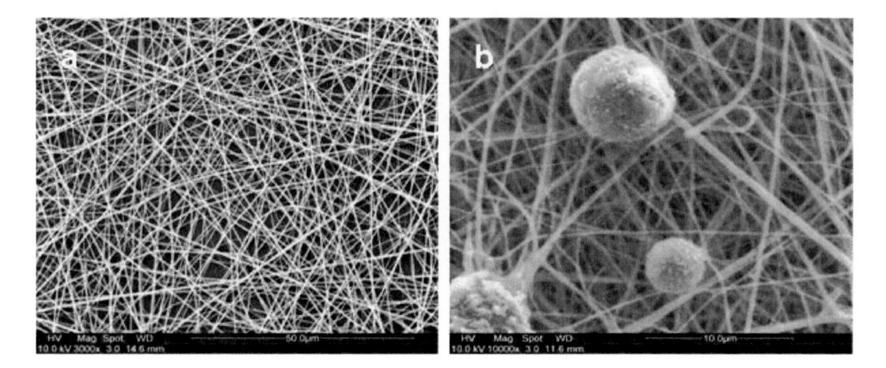

Figure 8. SEM images of (a) PCL/Collagen nanofibers (276 nm), (b) Umbilical cord blood hematopoietic stem cells on biocomposite PCL/collagen nanofibers.

Polyamide 3D fibrillar surfaces were utilized by Kamal *et al.* [103], for the self-renewal of mouse ESCs through mechanism involving Rac and P13K/AKT signaling, showing the role of nanostructures in maintaining stemness in ESC proliferations. Optimization of a suitable microenvironment needed for efficient differentiation of embryonic stem cells in 3D scaffold structures also involved the studies on scaffold pore size, increasing mechanical stiffness, increasing the cell seeding density, co-culturing with stromal cells etc. [104, 105]. The clinical application of stem cells shows evidence on the feasibility, efficacy and safety of MSCs in various medical conditions such as myocardial infraction, osteogenesis imperfecta. Hence, with a better understanding on the properties and behavior of MSCs on electrospun nanofibers, a 'stem cell-scaffold constructs' might serve as powerful tools for regenerative medicine curing various human disorders in the near future.

Conclusions

Nanotechnology and stem cells are exciting multidisciplinary research that promises to bring exciting discoveries over time in the field of regenerative medicine. Electrospinning offers a rapid, cost-effective and convenient way for mass production of nanofibers for fabricating scaffolds with biomolecules and has been utilized across a broad range of biocomposite polymer systems and tissue engineering endeavors. The need to improve the biomechanical properties and cell binding sites of electrospun nanofibers is paramount and also major obstacle currently facing tissue engineers. Nanotechnology will play a key role in medicine for providing revolutionary opportunities for early disease detection, diagnostic and therapeutic procedures to improve health and enhancing human physical abilities, enabling precise and effective therapy tailored to patients. The potentials of electrospinning as a means to create effective ECM analogue nanofibrous scaffolds is not limited to the selected tissues listed here, and there are numerous other possibilities, including, cardiac, cartilage, ligament, corneal and esophageal tissues. The future of nanomedicine will depend on rational design of bionanomaterials and tools based around a detailed and thorough understanding of biological systems rather than forcing applications for some materials currently in vogue.

ACKNOWLEDGMENTS

This study was supported by the Office of Life Sciences, National University of Singapore and StemLife Sdn BhD, Kuala Lumpur, Malaysia.

REFERENCES

[1] Venugopal, J., Sharon Low., Aw Tar Choon., and Ramakrishna, S. (2007). Interaction of Cells and Nanofiber scaffolds in Tissue engineering, *J. Biomed. Mater. Res. B*, 84, 34-48.

[2] Li, W.J., Laurencin, C.T, and Ko, F.K. (2002). Electrospun nanofibrous structure: A novel scaffold for tissue engineering, *J. Biomed. Mater. Res,* 60, 613-621.

[3] Venugopal, J., Ma, L.L., Yong, T., and Ramakrishna, S. (2005). In vitro study of smooth muscle cells on polycaprolactone and collagen nanofibrous matrices, *Cell Biol. Int,* 29, 861-867.

[4] Bleiziffer, O., Eriksson, E., and Kneser, U. (2007). Gene transfer strategies in tissue engineering, *J. Cell Mol. Med*, 11, 206-223.

[5] Huang, Z.M., Zhang, Y.Z., Kotaki, M., and Ramakrishna S. (2003). A review of polymer nanofibers by electrospinning and their applications in nanocomposites, *Comp. Sci. Tech,* 63, 2223-2253.

[6] Zhang, Y.Z., Lim, C.T., Ramakrishna, S., and Huang, Z.M. (2005). Recent development of polymer nanofibers for biomedical and biotechnological applications, *J. Mater. Sci. Mater. Med,* 16, 933-946.
[7] Venugopal, J., Zhang, Y.Z., and Ramakrishna, S. (2005). Fabrication of modified and functionalized polycaprolactone nanofiber scaffolds for vascular tissue engineering, *Nanotechnology,* 16, 2138-2142.
[8] Zhang, Y.Z., Ouyang, H.W., Lim, C.T., Ramakrishna, S., and Huang, Z.M. (2005). Electrospinning of gelatin fibers and Gelatin/PCL composite fibrous scaffolds, *J. Biomed. Mater. Res. B,* 72, 156-165.
[9] Badami, A.S., Kreke, M.R., and Goldstein, A.S. (2006). Effect of fiber diameter on spreading, proliferation, and differentiation of osteoblastic cells on electrospun poly(lactic acid) substrates, *Biomaterials,* 27, 596-606.
[10] Ma, P.X., Langer, R., In: M Yarmush., J. Morgan. Fabrication of biodegradable polymer foams for cell transplantation and tissue engineering. *Eds*, *Tissue engineering Methods and Protocols*, Humana Press, Inc, NJ, 1999, pp. 47-56.
[11] Ma, P.X., and Choi, J.W. (2001). Biodegradable polymer scaffolds with well defined interconnected spherical pore network, *Tissue Eng,* 7, 23-33.
[12] Zhang, Y.Z., Wang, X., Feng, Y., Li, J., Lim, C.T., and Ramakrishna, S. (2006). Coaxial electrospinning of (fluorescein isothiocyanate-conjugated bovine serum albumin)-encapsulated poly(epsilon-caprolactone) nanofibers for sustained release, *Biomacromolecules,* 7, 1049-1057.
[13] Reneker, D.H., and Chun, I. (1996). Nanometre diameter fibres of polymer produced by electrospinning, *Nanotechnology,* 7, 216-223.
[14] Inai, R., Kotaki, M., and Ramakrishna, S. (2005). Structure and properties of electrospun PLLA single nanofibres, *Nanotechnology,* 16, 208-213.
[15] Zhang, Y.Z., Huang, Z.M., Lim, C.T., and Ramakrishna, S. (2004). Preparation of core-shell structured PCL-r-gelatin bi-component nanofibers by coaxial electrospinning, *Chem. Mater,* 16, 3406-3409.
[16] Zhang, Y.Z., Venugopal, J., Huang, Z.M., Lim, C.T., and Ramakrishna, S. (2005). Characterization of the surface biocompatibility of the electrospun PCL-Collagen nanofibers using fibroblasts, *Biomacromolecules,* 6, 2583-2589.
[17] Jiang, H., Hu, Y., Li, Y., Zhao, P., Zhu, K., and Chen, W. (2005). A facile technique to prepare biodegradable coaxial electrospun nanofibers

for controlled release of bioactive agents, *J. Controlled Release,* 108, 237–243.

[18] Zhao, Y., and Cao, X. (2007). "Bio-mimic Multichannel Microtubes by a Facile Method," *J. Am. Chem. Soc,* 129, 764-765.

[19] Langer, R., and Vacanti, J.P. (1993). Tissue Engineering. *Science*, 260, 920-926.

[20] Chapekar, M.S. (2000). Tissue engineering: challenges and opportunities, *J. Biomed. Mater. Res,* 53, 617-620.

[21] Smith, L.A., and Ma, P.X. (2004). Nano-fibrous scaffolds for tissue engineering. Colloids Surf B, *Biointerfaces*, 39, 125-131.

[22] Kidoaki, S., Kwon, K., and Matsuda, T. (2005). Mesoscopic spatial designs of nano- and micro meshes for tissue-engineering matrix and scaffold based on newly devised multilayering and mixing electrospinning techniques, *Biomaterials,* 26, 37-46.

[23] Boateng, J.S., Matthews, K.H., Stevens, H.N.E., and Eccleston, G.M. (2008). Wound healing dressings and drug delivery systems: A review, *J. Pharm. Sciences,* 97, 2892-2923.

[24] Venugopal, J., Ma, L.L., and Ramakrishna, S. (2005). Biocompatible nanofiber matrices for engineering dermal substitute for skin regeneration, *Tissue Eng,* 11, 847-854.

[25] Venugopal, J., Zhang, Y.Z., and Ramakrishna, S. (2006). In vitro culture of human dermal fibroblasts on electrospun polycaprolactone collagen nanofibrous membrane, *Artif. Organs,* 30, 438-444.

[26] Sell, S., Barnes C., Smith, M., McCure, M., and Bowlin G. (2007). Extracellular matrix generated: tissue engineering via electrospun biomimetic nanofibers, *Polym. Int,* 56, 1349-60.

[27] Stevenson, S., Emery, S.E. and Goldberg, V.M. (1996). Factors Affecting Bone Graft Incorporation, *Clin. Ortho. Rel. Res*, 324: 66–74.

[28] Goldberg, V.M., and Stevenson, S. (1994). Bone graft options: fact and fancy, *Orthopedic*s, 17: 809–810, 821.

[29] Rose, F.R., and Oreffo, R.O. (2002). Bone tissue engineering: hope vs hype, *Biochem. Biophys. Res. Commun,* 292, 1-7.

[30] Bell, R., and Beirne, O.R. (1988). Effect of hydroxyapatite, tricalcium phosphate, and collagen on the healing of defects in the rat mandible, *J. Oral Maxillofac Surg,* 46, 589-594.

[31] Mehlisch, D.R., Taylor, T.D., Leibold, D.G., and Koretz, M.M. (1987). Evaluation of collagen/hydroxyapatite for augmenting deficient alveolar ridges: a preliminary report, *J. Oral Maxillofac Surg*, 1987, 45, 408-413.

[32] Ameer, G.A., Mahmood, TA., and Langer, R. (2002). A biodegradable composite scaffold for cell transplantation, *J. Ortho. Res.*, 20, 16-19.
[33] Hong, Z., Zhang P., He, C., and Jing, X. (2005). Nano-composite of poly(l-lactide) and surface grafted hydroxyapatite: Mechanical properties and biocompatibility, *Biomaterials*, 26, 6296-6304.
[34] Joshi, R.R., Underwood, T., Frautschi, J.R., and Levy, R.J. (1996). Calcification of polyurethanes implanted subdermally in rats is enhanced by calciphylaxis, *J. Biomed. Mater. Res,* 31, 201-207.
[35] Liu, X., Smith, L.A., Hu, J., and Ma, P.X. (2009). Biomimetic nanofibrous gelatin/apatite composite scaffolds for bone tissue engineering, *Biomaterial*s, 30, 2252-2258.
[36] Yoshimoto, H., Shin, Y.M., Terai, H., and Vacanti, J.P. (2003). A biodegradable nanofiber scaffold by electrospinning and its potential for bone tissue engineering, *Biomaterials*, 24, 2077-2082.
[37] Walker-Simmons, M., and Ryan, C.A. (1984). Induction by Chitosan Oligomers and Chemically Modified Chitosan and Chitin, *Plant Physiol,* 76, 787-790.
[38] Hu, Q., Li, B., Wang, M., and Shen, J. (2004). Preparation and characterization of biodegradable chitosan/hydroxyapatite nanocomposite rods via in situ hybridization: a potential material as internal fixation of bone fracture, *Biomaterials*, 25, 779–785.
[39] Zhang, Y., Venugopal, J.R., El-Turki, A., Ramakrishna, S., Su, B., and Lim, C.T. (2008). Electrospun biomimetic nanocomposite nanofibers of hydroxyapatite/chitosan for bone tissue engineering, *Biomaterials*, 2008, 29, 4314-4322.
[40] Liao, S.S., Cui, F.Z., and Feng, Q.L. (2004). Hierarchically biomimetic bone scaffold materials: nano-HA/collagen/PLA composite, *J. Biomed. Mater. Res B,* 69, 158-165.
[41] Venugopal, J., Vadgama, P., and Ramakrishna, S. (2007). Biocomposite nanofibers and osteoblasts for bone tissue engineering, *Nanotechnology,* 18, 511-518.
[42] Gupta, D., Venugopal, J., Mitra, S., Giri Dev, V.R., and Ramakrishna, S. (2009). Nanostructured biocomposite substrates by electrospinning and electrospraying for the mineralization of osteoblasts, *Biomaterials*, 2009, 30, 2085-3094.
[43] Venugopal, J., Sharon Low., Aw Tar Choon., and Ramakrishna, S. (2008). Nanobioengineered Electrospun Composite Nanofibers and Osteoblasts for Bone Regeneration, *Artif. Organs*, 32, 388-397.

[44] Itoh, S., Kikuchi, M., and Shinomiya, K. (2001). The biocompatibility of osteoconductive activity of a novel hydroxyapatite/collagen composite biomaterial, and its function as a carrier of rhBMP-2, *J. Biomed. Mater. Res,* 54, 445-453.

[45] Kikuchi, M., Ikoma, T., Shinomiya, K., and Tanaka, J. (2004). Biomimetic synthesis of bone-like nanocomposites using the self-organization mechanism of hydroxyapatite and collagen, *Com. Sci. Tech,* 64, 819-825.

[46] Nisbet, D.R., Forsythe, J.S., Shen, W., Finkelstein, D.I., and Horne, M.K. (2008) A Review of the Cellular Response on Electrospun Nanofibers for Tissue Engineering, *J. Biomater. Appl,* December 12.

[47] Rodengen, J.L., Stephensen, L.W. Coronary artery diseases and treatment options. In: Rodengen JL, Stephensen, LW, editors. The practical Guide to Your Hear and Heart Surgery. Fort Lauderdale, FL: Write Stuff Enterprises Inc.; 1999. pp. 110-129.

[48] Cummings, C.L., Gawlitta, D., Nerem, R.M., and Stegemann, J.P. (2004). Properties of engineered vascular constructs made from collagen, fibrin, and collagen-fibrin mixtures, *Biomaterials*, 25, 3699-3706.

[49] Isenberg, B.C., Williams, C., and Tranquillo, R.T. (2006). Small-Diameter Artificial Arteries Engineered In Vitro, *Cir. Res,* 98, 25-35.

[50] He, W., Yong, T., Ma, Z., Teo, W.E., and Ramakrishna, S. (2005). Fabrication and endothelialization of collagen-blended biodegradable polymer nanofibers: potential vascular graft for blood vessel tissue engineering, *Tissue Eng,* 11, 1574-1588.

[51] Mo, X.M., Xu, C.Y., Kotaki, M., and Ramakrishna, S. (2004). Electrospun P(LLA-CL) nanofiber: a biomimetic extracellular matrix for smooth muscle cell and endothelial cell proliferation, *Biomaterials,* 25, 1883-1890.

[52] Shum-Tim, D., Stock, U., and Moses, M.A. (1999). Tissue engineering of autologous aorta using a new biodegradable polymer, *Ann. Thorac Surg,* 68, 2298-2305.

[53] Shin'oka, T., Imai, Y., and Ikada, Y. (2001). Transplantation of a tissue engineered pulmonary artery, *New Engl. J. Med,* 344,532-533.

[54] Matsumura, G., Hibino, N., and Shin'oka, T. (2003). Successful application of tissue engineered vascular autografts: clinical experience, *Biomaterials,* 24, 2303-2308.

[55] Kakisis, J.D., Liapis, C.D., Breuer, C., and Sumpio, B.E. (2005). Artificial blood vessel: The Holy Grail of peripheral vascular surgery, *J. Vasc. Surg,* 41, 349-354.
[56] Theron, A., Zussman, E., and Yarin, A.L. (2001). Electrostatic field-assisted alignment of electrospun nanofibres, *Nanotechnology,* 12, 384-390.
[57] Xu, C.Y., Inai, R., Kotaki, M., and Ramakrishna, S. (2004). Aligned biodegradable nanofibrous structure: a potential scaffold for blood vessel engineering, *Biomaterials,* 25, 877-886.
[58] Xu, C., Inai, R., Kotaki, M., and Ramakrishna, S. (2004). Electrospun nanofibers fabrication as synthetic extracellular matrix and its potential for vascular tissue engineering, *Tissue Eng,* 10, 1160-1168.
[59] He Wei., Yong T., Teo, W.E., and Ramakrishna, S. (2006). Biodegradable polymer nanofiber mesh to maintain functions of endothelial cells, *Tissue Eng,* 12, 2457-2466.
[60] Stoll, G., and Muller, H.W. (1999). Nerve injury, axonal degeneration and neural regeneration: basic insights, *Brain Pathol*, 9, 313-25.
[61] Navarro, X., Rodriguez F.J., Ceballos, D., and Verdu, E. (2003). Engineering an artificial nerve graft for the repair of severe nerve injuries. *Med. Biol. Eng. Comp*, 41, 220-226.
[62] Yu, X., and Bellamkonda, R.V. (2003). Tissue engineered scaffolds are effective alternatives to autografts for bridging peripheral nerve gaps, *Tissue Eng,* 9(3), 421-430.
[63] Bini, T.B., Gao, S., and Tan, T.C. (2004). Electrospun poly(L-lactide-co-glycolide) biodegradable polymer nanofiber tubes for peripheral nerve regeneration, *Nanotechnology*, 15(11): 1459-1464.
[64] Koh, H.S., Thomas, Y., Chan C.K., and Ramakrishna, S. (2008). Enhancement of neurite outgrowth using nano-structured scaffolds coupled with laminin. *Biomaterials*, 29, 3574-82.
[65] Powell, M.P., Sobarzo, M.R., and Saltzman, W.M. (1990). Controlled Release of Nerve Growth Factor from a Polymeric Implant, *Brain Res,*. 515, 309-311.
[66] Aebischer, P., Salessiotis, A.N., and Winn, S.R. (1989). Basic fibroblast growth factor released from synthetic guidance channels facilitates peripheral nerve regeneration across long gaps, *J. Neurosci. Res,* 23, 282-289.
[67] Cordeiro, P.G; (1989). Acidic fibroblast growth factor enhances peripheral nerve regeneration in vivo, *Plast. Reconstr. Surg,* 83, 1013-1019.

[68] Yang, F., Murugan, R., Wang, S., and Ramakrishna, S. (2005). Electrospinning of nano/micro scale poly(L-lactic acid) aligned fibers and their potential in neural tissue engineering, *Biomaterials,* 26(15), 2603- 2610.

[69] Ghasemi, L,M., Prabhakaran, M.P., Morshed, M., Nasr-Esfahani, M.H., and Ramakrishna, S. (2008). Electrospun poly(ε-caprolactone)/gelatin nanofibrous scaffolds for nerve tissue engineering, *Biomaterials*, 29(34), 4532-4539.

[70] Schnell, E., Kinkhammer, K., Balzer, S., Brook, G., and Mey, J. (2007). Guidance of glial cell migration and axonal growth on electrospun nanofibers of poly-ε-caprolactone and a collagen/poly-ε-caprolactone blend, *Biomaterials*, 28, 3012–25.

[71] Madison, R., da Silva, C.F., Dikkes, P., and Sidman, R.L. (1985). Increased rate of peripheral nerve regeneration using bioresorbable nerve guides and a laminin-containing gel, *Exp. Neurol,* 88, 767-772.

[72] Chen, Y.S., Hsieh, C.L., Tsai, C.C., Chen T.H., and; Yao, C.H. (2000). Peripheral nerve regeneration using silicone rubber chambers filled with collagen, laminin and fibronectin, *Biomaterials,* 21, 1541-1547.

[73] Chew, S.Y., Mi, R., Leong, K.W., and Hoke, A. (2008). The effect of the alignment of electrospun fibrous scaffolds on Schwann cell maturation, *Biomaterials*, 29, 653- 661.

[74] Yuan, Y., Zhang, P., Yang, Y., Wang, X., and Gu, X. (2004). The interaction of Schwann cells with chitosan membranes and fibers in vitro, *Biomaterials*, 25, 4273-4278.

[75] Prabhakaran, M.P., Venugopal, J., Chyan, T.T., Hai, L.B., Chan, C.K., Tang, A.L., and Ramakrishna, S. (2008). Electrospun biocomposite nanofibrous scaffolds for neural tissue engineering, *Tissue Eng. A*, 14(11), 1787-1797.

[76] Prabhakaran MP, Venugopal, J., Casey Chan., and Ramakrishna S. (2008). Surface modified electrospun nanofibrous scaffolds for nerve tissue engineering, *Nanotechnology*, 19, 455102 (8pp).

[77] Gupta, D., Venugopal, J., Prabhakaran, M.P., Giri Dev, V.R., and Ramakrishna, S. (2009). Aligned and random nanofibrous substrate for the in vitro culture of Schwann cells for neural tissue engineering, *Acta Biomater*, online.

[78] Guenard, V., Kleitman, N., Morrissey, T.K., and Aebischer, P. (1992). Syngeneic Schwann cells derived from adult nerve seeded in semipermeable guidance channels enhance peripheral nerve regeneration, *J. Neurosci,* 12, 3310-3316.

[79] Ide C. (1996). Peripheral nerve regeneration, *Neuroci Res*, 25,101–121.
[80] Hall S. (2001). Nerve repair: a neurobiologist's view, *J. Hand Surg*, 26B, 129-136.
[81] Ansselin, A.D., Fink, T., and Davey, D.F. (1997). Peripheral nerve regeneration through nerve guides seeded with adult Schwann cells, *Neuropathol. Appl. Neurobiol*, 23, 387-398.
[82] Whitworth, I.H., Brown, R.A., Dore, C., Green, C.J., and Terenghi, G. (1995). Orientated mats of fibronectin as a conduit material for use in peripheral nerve repair, *J. Hand Surg. J. British Soc. Surg. hand*, 20, 429-436.
[83] Miller, C., Shanks, H., Witt, A., Rutkowski, G., and Mallapragada, S. (2001). Oriented Schwann cell growth on micropatterened biodegradable polymer substrates, *Biomaterials*, 22(11), 1263-1269.
[84] Borkenhagen, M., Clemence, J.F., Sigrist, H., and Aebischer, P. (1998). Three dimensional extracellular matrix engineering in the nervous system, *J. Biomed. Mater Res*, 40(3): 392-400.
[85] Mannello, F., and Tonti, G.A. (2007). Concise Review: No Breakthroughs for Human Mesenchymal and Embryonic Stem Cell Culture: Conditioned Medium, Feeder Layer, or Feeder-Free; Medium with Fetal Calf Serum, Human Serum, or Enriched Plasma; Serum-Free, Serum Replacement Nonconditioned Medium, or Ad Hoc Formula? All That Glitters Is Not Gold! *Stem Cells,* 25, 1603–1609.
[86] Phinney, D.G., and Prockop, D.J. (2007). Concise Review: Mesenchymal Stem/ Multipotent Stromal Cells: The State of Transdifferentiation and Modes of Tissue Repair-Current Views, *Stem Cells*, 25, 2896-2902.
[87] Hui, J.H.P., Ouyang, H. W., and Hutmacher, D.W. (2005). Mesenchymal stem cells in muscoskeletal tissue engineering: A review of recent advances in National University of Singapore, *Ann. Acad. Med. Singapore*, 34, 206-212.
[88] Cui, L., Jiang, J., Wei, L., Zhou, X., Fraser, J.L., and Snider, B.J. (2007). Transplantation of embryonic stem cells improves nerve repair and functional recovery after severe sciatic nerve anatomy in rats, *FASEB J*, 21, 1–11.
[89] Cui, L., Jiang, J., Wei, L., Zhou, X ., and Snider, B.J. (2008). Transplantation of embryonic stem cells improves nerve repair and functional recovery after severe sciatic nerve anatomy in rats, *Stem Cells*, 26(5):1356–1365.

[90] Willerth, S.M., Rader, A., and Sakiyama-Elbert, S.E. (2008). The effect of controlled growth factor delivery on embryonic stem cell differentiation inside fibrin scaffolds, *Stem Cell Res*, 1(3), 205–18.

[91] Xie, J., Willerth, S.M., Li, X., Macewan, M.R., Rader, A., and Xia, Y. (2009). The differentiation of embryonic stem cells seeded on electrospun nanofibers into neural lineages, *Biomaterials*, 30, 354–362.

[92] Kazemnejad, S., Allameh, A., Soleimani, M., Gharehbaghian, A., and Jazayery, M. (2009). Biochemical and molecular characterization of hepatocyte-like cells derived from human bone marrow mesenchymal stem cells on a novel three-dimensional biocompatible nanofibrous scaffold, *J. Gastroent Hepatol*, 24(2), 278-287.

[93] Schofer, M.K., Boudriot, U., Wack, C., Leifeld, I., Grabedunkel, C., and Winkelmann, S.F. (2009). Influence of nanofibers on the growth and osteogenic differentiation of stem cells: a comparison of biological collagen nanofibers and synthetic PLLA fibers, *J. Mater. Sci. Mater. Med*, 20(3), 767-774.

[94] Makino, S., Fukuda, K., and Miyoshi, S. (1999). Cardiomyocytes can be generated from marrow stromal cells in vitro, *J Clin Invest*, 103, 697–705.

[95] Balana, B., Nicoletti, C., and Zahanich, I. (2006). 5-Azacytidine induces changes in electrophysiological properties of human mesenchymal stem cells, *Cell Res*, 16, 949 –990.

[96] Davis, M.E., Hsieh, P.C., Grodzinsky, A.J., and Lee, R.T. (2005). Custom design of the cardiac microenvironment with biomaterials, *Circ. Res,* 97, 8-15.

[97] Li, W.J., Tuli, R., Huang, X., Laquerriere, P., and Tuan R.S. (2005). Multi lineage differentiation of human mesenchymal stem cells in a three dimensional nanofibrous scaffold, *Biomaterials*, 26 (25), 5158-5166.

[98] Chivu, M., Diaconu, C.C., Bleotu, C., and Cernescu, C. (2004). The comparison of different protocols for expansion of umbilical-cord blood hematopoietic stem cells, *J. Cell Mol. Med*, 8(2), 223-231.

[99] Tabbera, I.A. (2002). Allogeneic heamatopoietic stem cell transplantation: complications and results, *Ach. Inter. Med*, 162, 1558-1566.

[100] De Vries, E.G.E., Vellenga, E., Kluin-Nelemans, J.C., and Mulder, N.H. (2004). The happy density of frozen haemotopoietic stem cells: from immature stem cells to mature applications, *Eur. J. cancer,* 40, 1987-1992.

[101] Chua, K.N., Chai, C., Lee, P.C., Tang, Y.N., Ramakrishna, S., Leong, K.W., Mao, H.Q. (2006). Surface-aminated electrospun nanofibers enhance adhesion and expansion of human umbilical cord blood hematopoietic stem/progenitor cells, *Biomaterials*, 27, 6043-6051.

[102] Chua, K.N., Chai. C., Lee, P.C., Ramakrishna, S., Leong, K.W., and Mao, H.Q. (2007). Functional nanofiber scaffolds with different spacers modulate adhesion and expansion of cryopreserved umbilical cord blood hematopoietic stem/progenitor cells, *Exp. Hematol*, 35, 771-781.

[103] Kamal, A.N., Ahmed, I., Kamal, J., Schindler, M.N., and Meiners, S. (2006). Three-Dimensional nanofibrillar surfaces promote self-renewal in mouse embryonic stem cells, *Stem Cells*, 24, 426-433.

[104] Taqvi, S., and Roy, K. (2006). Influence of scaffold physical properties and stromal cell coculture on hematopoietic differentiation of mouse embryonic stem cells. *Biomaterials*, 27, 6024–6031.

[105] Gauthaman, K., Venugopal, J.R., Yee, F.C., Peh, G.S., Ramakrishna, S., and Bongso, A. (2009). Nanofibrous substrates support colony formation and maintain stemness of human embryonic stem cells, *J. Cell Mol. Med*, 2009, Feb 17.

In: Electrospinning Process and Nanofiber… ISBN 978-1-61209-330-7
Editors: A.K. Haghi and G.E. Zaikov

Chapter 6

MULTILAYERS ELECTROSPUN NANOFIBER WEB: A NEW CLASS OF NONWOVENS

A. K. Haghi[*]
University of Guilan, Rasht, Iran

ABSTRACT

Electrospun nanofiber web has many potential applications due to its large specific area, very small pore size and high porosity. Despite such potentials, the mechanical properties of nanofiber web are very poor for use in textile application.

To remedy this defect, laminating process could accomplish in order to protect nanofiber web versus mechanical stresses. But in these processes nanofiber properties may change during the process. In the first part of this article the influence of laminating temperature on nanofiber/laminate properties is studied. In the second part, a simulation algorithm has been employed for generating nonwovens with known characteristics.

Since the physical characteristics of simulated images are known exactly, one can employ them to test the usefulness of algorithm used in characterizing diameter and other structural features.

Keywords:Electrospun nanofiber, Laminating, simulation.

---

[*] Haghi@Guilan.ac.ir

Introduction

Nowadays, there are different types of protective clothing that some of these are disposable and non-disposable. The simplest and most preliminary of this equipment is made from rubber or plastic that is completely impervious to hazardous substances. Unfortunately, these materials are also impervious to air and water vapor, and thus retain body heat, exposing their wearer to heat stress which can build quite rapidly to a dangerous level. Another approach to protective clothing is incorporating activated carbon into multilayer fabric in order to absorb toxic vapors from environment and prevent penetration to the skin. The use of activated carbon is considered only a short term solution because it loses its effectiveness upon exposure to sweat and moisture. The use of semi-permeable membranes as a constituent of the protective material is a another approach. In this way, reactive chemical decontaminants encapsulates in microparticles or fills in Microporous Hollow fibers and then coats onto fabric. The microparticle or fiber walls are permeable to toxic vapors, but impermeable to decontaminants, so that the toxic agents diffuse selectively into them and neutralize [1-3].

Generally, a negative relationship always exists between thermal comfort and protection performance for currently available protective clothing. Thus there still exists a very real demand for improved protective clothing that can offer acceptable levels of impermeability to highly toxic pollutions of low molecular weight, while minimizing wearer discomfort and heat stress.

Electrospinning provides an ultrathin membrane-like web of extremely fine fibers with very small pore size and high porosity, which makes them excellent candidates for use in filtration, membrane, and possibly protective clothing applications. Preliminary investigations have indicated that the using of nanofiber web in protective clothing structure could present minimal impedance to air permeability and extremely efficiency in trapping dust and aerosol particles. Meanwhile, it is found that the electrospun webs of nylon 6,6, polybenzimidazole, polyacrylonitrile, and polyurethane provided good aerosol particle protection, without a considerable change in moisture vapor transport or breathability of the system. While nanofiber webs suggest exciting characteristics, it has been reported that they have limited mechanical properties. In order to provide suitable mechanical properties for use as cloth, nanofiber webs must be laminated via an adhesive into a fabric system. This system could protect ultrathin nanofiber web versus mechanical stresses over an extended period of time [4-6].

The adhesives could be as melt adhesive or solvent-based adhesive. When a melt adhesive is used, the hot-press laminating carried out at temperatures above the softening or melting point of adhesive. If a solvent-based adhesive is used, laminating process could perform at room temperature. In addition, the Solvent-based adhesive is generally environmentally unfriendly, more expensive and usually flammable, whereas the hot-melt adhesive is environmentally friendly, inexpensive requires less heat, and so is now more preferred. However, without disclosure of laminating details, the hot-press method is more suitable for nanofiber web lamination. In this method, laminating temperature is one of the most important parameters. Incorrect selection of this parameter may lead to change or damage nanofiber web. Thus, it is necessary to find out a laminating temperature which has the least effect on the nanofiber web.

It has been found that morphology such as fiber diameter and its uniformity of the electrospun polymer fibers are dependent on many processing parameters. These parameters can be divided into three groups as shown in Table 1. Under certain condition, not only uniform fibers but also beads-like formed fibers can be produced by electrospinning. Although the parameters of the electrospinning process have been well analyzed in each of polymers these information has been inadequate enough to support the electrospinning of ultra-fine nanometer scale polymer fibers. A more systematic parametric study is hence required to investigate.

Table 1. Processing parameters in electrospinning

Solution properties	Viscosity
	Polymer concentration
	Molecular weight of polymer
	Electrical conductivity
	Elasticity
	Surface tension
Processing conditions	Applied voltage
	Distance from needle to collector
	Volume feed rate
	Needle diameter
Ambient conditions	Temperature
	Humidity
	Atmospheric pressure

The purpose of this study is to consider the influence of laminating temperature on nanofiber/laminate properties. Multilayer fabrics were made by electrospinnig polyacrylonitrile nanofibers onto nonwoven substrate and incorporating into fabric system via hot-press method at different temperatures.

EXPERIMENTAL

Electrospining and Laminating Process

Polyacrylonitrile (PAN) of 70,000 g/mol molecular weight from Polyacryl Co. (Isfehan, Iran) has been used with Dimethylformamide (DMF) from Merck, to form a polymer solution 12% w/w after stirring for 5h and exposing for 24h at ambient temperature. The yellow and ripen solution was inserted into a plastic syringe with a stainless steel nozzle 0.4 mm in inner diameter and then it was placed in a metering pump from WORLD PRECISION INSTRUMENTS (Florida, USA). Next, this set installed on a plate which it could traverse to left-right along drum (Figure 1). The flow rate 1 μl/h for solution was selected and the fibers were collected on an aluminum-covered rotating drum (with speed 9 m/min) which was previously covered with a polypropylene spun-bond nonwoven (PPSN) substrate of 28cm× 28cm dimensions; 0.19 mm thickness; 25 g/m2 weight; 824 $cm^3/s/cm^2$ air permeability and 140ºC melting point.

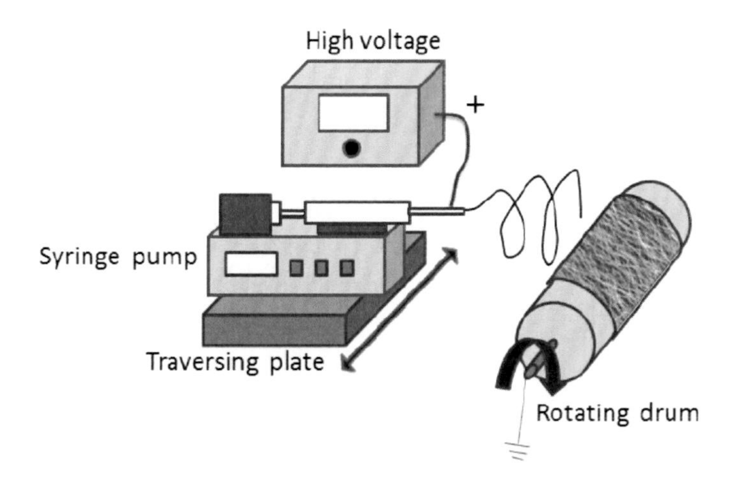

Figure1. Electrospinning setup.

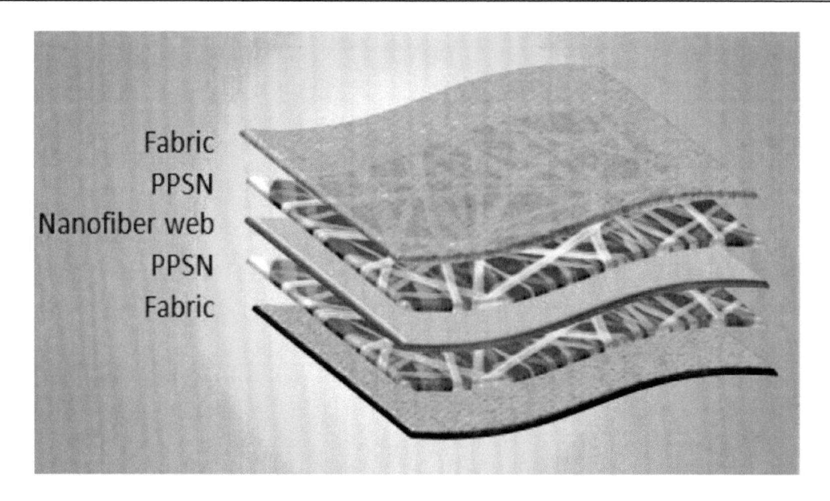

Figure 2. Multilayer fabric components.

The distance between the nozzle and the drum was 7cm and an electric voltage of approximately 11kV was applied between them. Eelectrospinning process was carried out for 8h at room temperature to reach approximately web thickness 3.82 g/m^2. Then nanofiber webs were laminated into cotton weft-warp fabric with a thickness 0.24mm and density of 25×25 (warp-weft) per centimeter to form a multilayer fabric (Figure 2). Laminating was performed at temperatures 85,110,120,140,150°C for 1 min under a pressure of 9 gf/cm^2.

Nanofiber Web Morphology

In order to consider of nanofiber web morphology after hot-pressing, another laminating was performed by a non-stick sheet made of Teflon (0.25 mm thickness) instead one of the fabrics (fabric /pp web/nanofiber web/pp web/non-stick sheet). Finally, after removing of Teflon sheet, the nanofiber layer side was observed under an optical microscope (MICROPHOT-FXA, Nikon, Japan) connected to a digital camera.

Measurement of Air Permeability

Air permeability of multilayer fabric after lamination was tested by TEXTEST FX3300 instrument (Zürich, Switzerland). It was tested 5 pieces of each sample under air pressure 125pa at ambient condition (16°C, 70%RH) and obtained average air permeability.

RESULTS AND DISCUSSION

PPSN was selected as melt adhesive layer for hot-press laminating (Figure 2). This process was performed under different temperatures to find an optimum condition.

Figure 3 presents the optical microscope images of nanofiber web after lamination. It is obvious that by increasing of laminating temperature to melting point (samples a-c) the adhesive layer gradually melts and spreads on web surface. But, when melting point selected as laminating temperature (sample d) the nanofiber web begin to damage.

In this case, the adhesive layer completely melted and penetrated into nanofiber web and occupied its pores. This procedure intensified by increasing of laminating temperature above melting point. As shown in Figure 1. (sample e), perfect absorption of adhesive by nanofiber web creates a transparent film which leads to appear fabric structure.

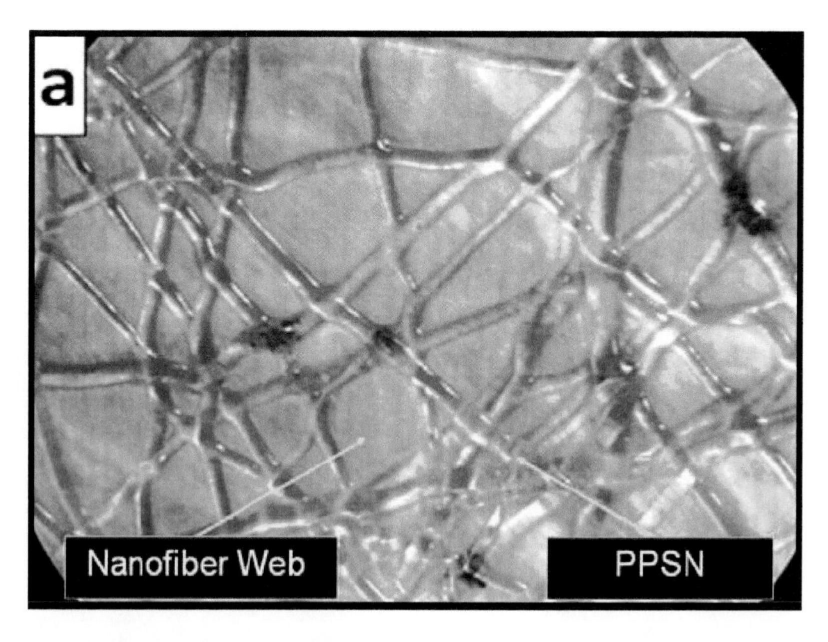

Figure 3. a. The optical microscope images of nanofiber web after laminating at 85°C (at 100 magnification).

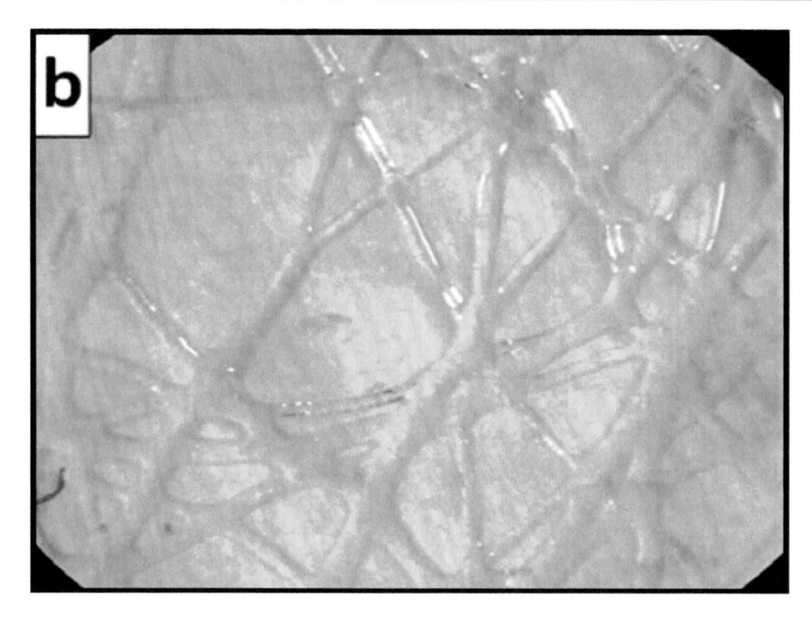

Figure 3.b. The optical microscope images of nanofiber web after laminating at 110°C (at 100 magnification).

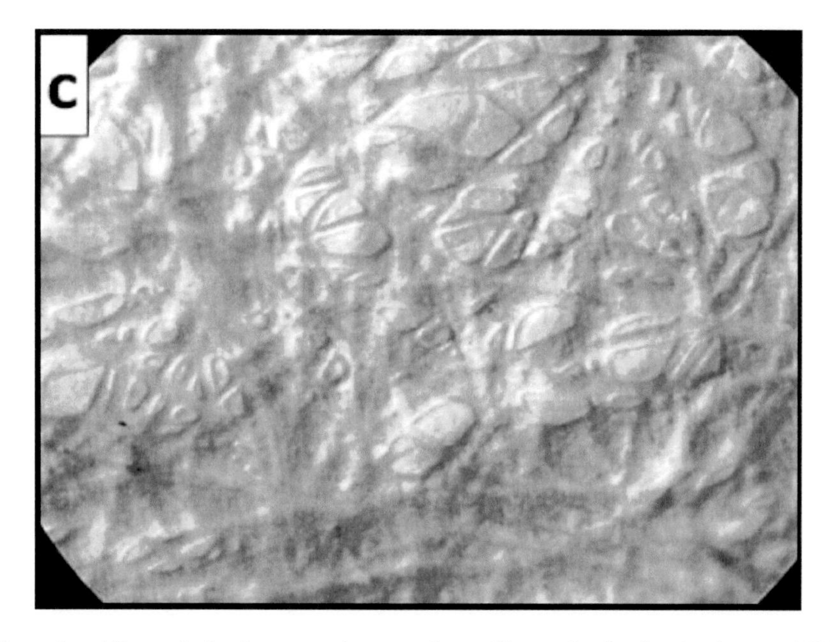

Figure 3. c. The optical microscope images of nanofiber web after laminating at 120°C (at 100 magnification).

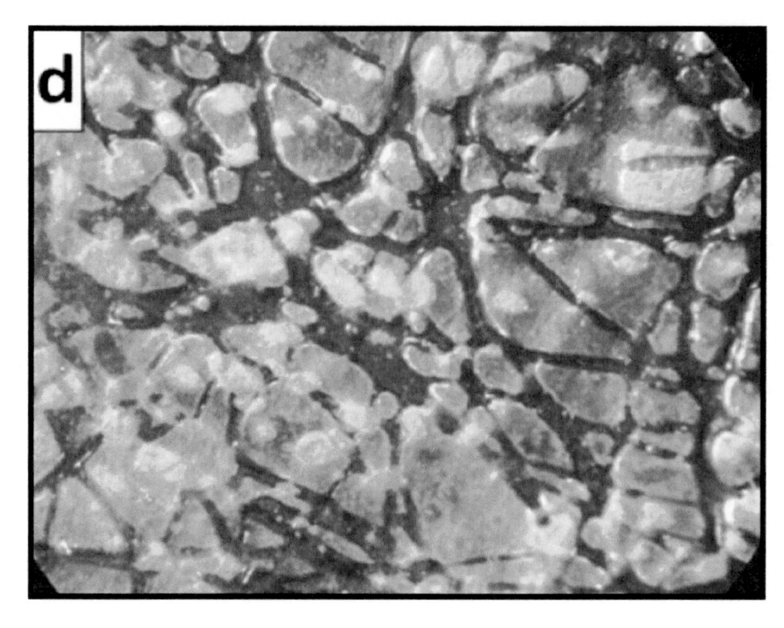

Figure 3.d. The optical microscope images of nanofiber web after laminating at 140°C (at 100 magnification).

Figure 3.e. The optical microscope images of nanofiber web after laminating at temperatures more than 140°C (at 100 magnification).

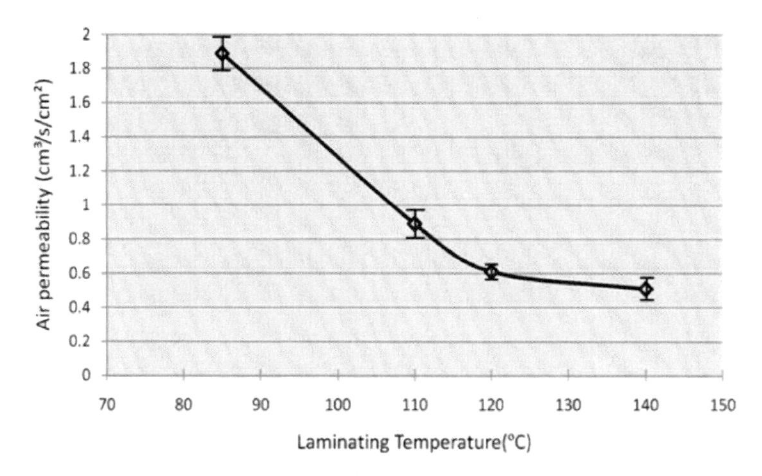

Figure 4. Air permeability of multilayer fabric as a function of laminating temperature.

Also, to examine how laminating temperature affect the breathability of multilayer fabric, air permeability experiment was performed. Figure 4. indicates the effect of laminating temperature on air permeability. As might be expected, air permeability decreased with increasing laminating temperature. This behavior attributed to melting procedure of adhesive layer. As mentioned above, before melting point the adhesive gradually spreads on web surface. This phenomenon causes that the adhesive layer act like an impervious barrier to air flow and reduces air permeability of multilayer fabric. But at melting point and above, the penetration of melt adhesive into nanofiber/fabric structure leads to fill its pores and finally decrease in air permeability. Furthermore, we only observed that the adhesive force between layers was increased according to temperature rise. The sample (a) exhibited very poor adhesion between nanofiber web and fabric and it could be separated by light abrasion of thumb, while adhesion increased by increasing laminating temperature to melting point. It must to be noted that after melting point because of passing of melt PPSN across nanofiber web, adhesion between two layers of fabric will occurred.

Mechanical Properties of Multilayer Nano-Web

The tensile strength of multilayer fabrics with and without nanofiber web, were carried out using MICRO250 tensile machine (SDL International Ltd.). Ten samples were cut from the warp directions of multilayer fabric at size of

10mm×200mm and then exposed to the standard condition (25°C,60% RH) for 24h in order to conditioning. To measure tensile strength, testing was performed by load cell of 25Kgf. Also, the distance between the jaws and the rate of extension were selected 100mm and 20mm/min, respectively .The tensile strength of samples without nanofibers (Figure 5) are weaker than those laminated with nanofibers (Figure 6). According to Table 2, the breaking load and breaking elongation for the samples laminated with electrospun nanofibers are improved as well. These variations can be observed clearly in Figures 7 and 8 for 10 samples.

Table 2. Tensile strength test results of the Multilayer fabrics

Multilayer Fabric	Warp direction			
	Breaking Load, N		Breaking Elongation, mm	
	Mean value	CV, %	Mean value	CV, %
Without Nanofiber web	174.427	6.2	5.02	7.5
With nanofiber web	189.211	4.6	5.11	6

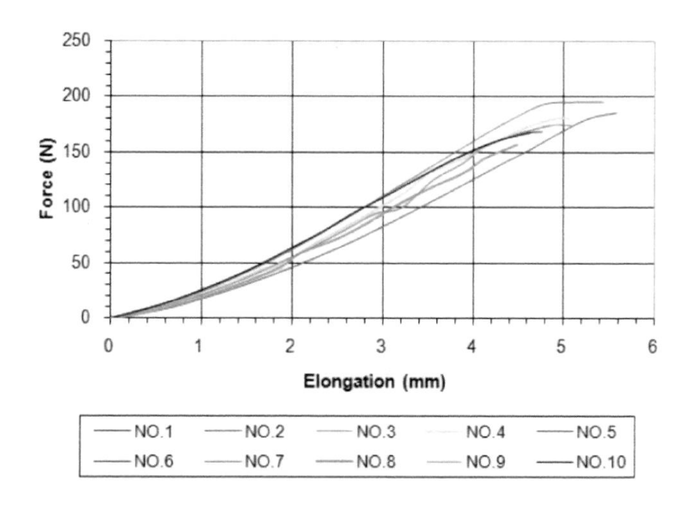

Figure 5. Force-Elongation curve for multilayer fabric without Nanofiber web.

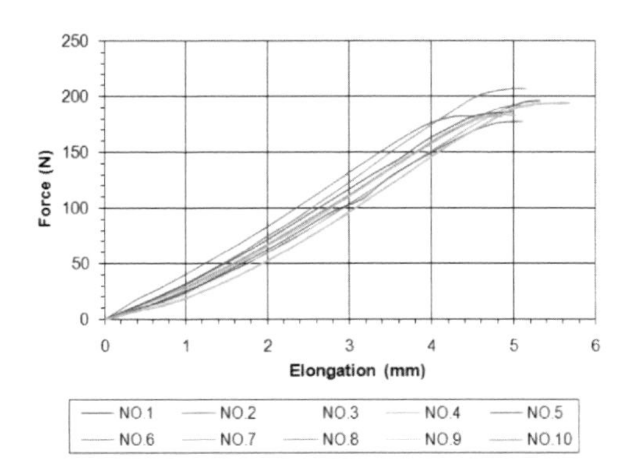

Figure 6. Force-Elongation curve for multilayer fabric with Nanofiber web.

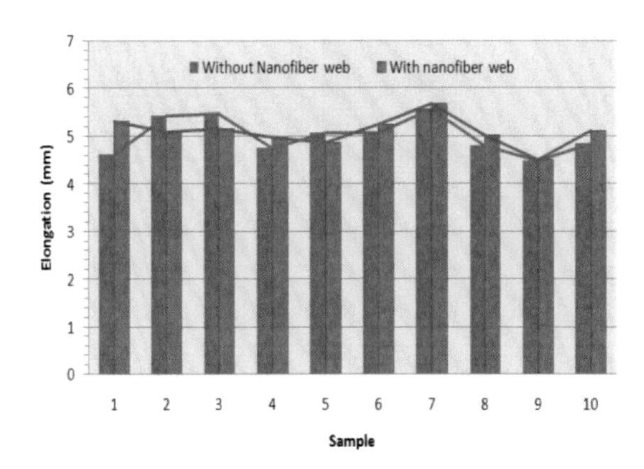

Figure 7. Breaking elongation of ten samples.

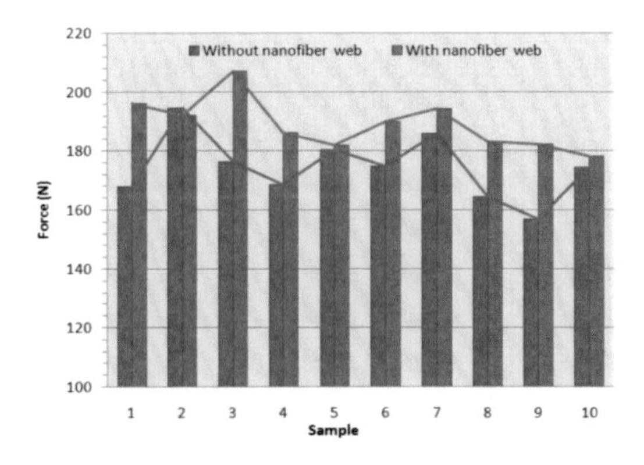

Figure 8. Breaking load of ten samples.

Simulation of Nano-Web

For the continuous fibers, it is assumed that the lines are infinitely long so that in the image plane, all lines intersect the boundaries. Under this scheme (Figure), a line with a specified thickness is defined by the perpendicular distance *d* from a fixed reference point *O* located in the center of the image and the angular position of the perpendicular α. Distance *d* is limited to the diagonal of the image. Based on the objective of this paper, several variables are allowed to be controlled during the simulation:

1. *Web density* that can be controlled using the line density which is the number of lines to be generated in the image.
2. *Angular density* which is useful for generating fibrous structures with specific orientation distribution. The orientation may be sampled from either a normal or a uniform random distribution.
3. *Distance from the reference point* normally varies between zero and the diagonal of the image, restricted by the boundary of the image and is sampled from a uniform random distribution.
4. *Line thickness* (fiber diameter) is sampled from a normal distribution. The mean diameter and its standard deviation are needed.
5. *Image size* can also be chosen as required.

Fiber Diameter Measurement

The first step in determining fiber diameter is to produce a high quality image of the web, called micrograph, at a suitable magnification using electron microscopy techniques. The methods for measuring electrospun fiber diameter are described in following sections.

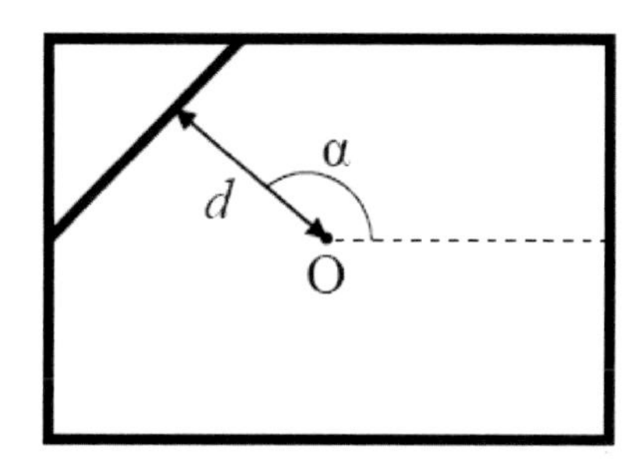

Figure 9. Procedure for μ-randomness.

Manual Method

The conventional method of measuring the fiber diameter of electrospun webs is to analyze the micrograph manually. The manual analysis usually consists determining the length of a pixel of the image (setting the scale), identifying the edges of the fibers in the image and counting the number of pixels between two edges of the fiber (the measurements are made perpendicular to the direction of fiber-axis), converting the number of pixels to *nm* using the scale and recording the result. Typically 100 measurements are carried out (Figure 10). However, this process is tedious and time-consuming especially for large number of samples. Furthermore, it cannot be used as on-line method for quality control since an operator is needed for performing the measurements. Thus, developing automated techniques which eliminate the use of operator and has the capability of being employed as on-line quality control is of great importance.

Distance Transform

The *distance transform* of a binary image is the distance from every pixel to the nearest nonzero-valued pixel. The center of an object in the distance transformed image will have the highest value and lie exactly over the object's *skeleton*. The skeleton of the object can be obtained by the process of *skeletonization* or *thinning*. The algorithm removes pixels on the boundaries of objects but does not allow objects to break apart. This reduces a thick object to its corresponding object with one pixel width. Skeletonization or thinning often produces short spurs which can be cleaned up automatically with a *pruning* procedure. The algorithm for determining fiber diameter uses a binary input image and creates its skeleton and distance transformed image. The skeleton acts as a guide for tracking the distance transformed image by recording the intensities to compute the diameter at all points along the skeleton. Figure 11. shows a simple simulated image, which consists of five fibers with diameters of 10, 13, 16, 19 and 21 pixels, together with its skeleton and distance map including the histogram of fiber diameter obtained by this method.

Direct Tracking

Direct tracking method uses a binary image as an input data to determines fiber diameter based on information acquired from two scans; first a horizontal and then a vertical scan. In the horizontal scan, the algorithm searches for the first white pixel adjacent to a black. Pixels are counted until reaching the first black. The second scan is then started from the mid point of horizontal scan and pixels are counted until the first black is encountered. Direction changes if the black pixel isn't found. Having the number of horizontal and vertical scans, the number of pixels in perpendicular direction which is the fiber diameter could be measured from a geometrical relationship. The explained process is illustrated in Figure 12. In electrospun nonwoven webs, nanofibers cross each other at intersection points and this brings about the possibility for some untrue measurements of fiber diameter in these regions. To circumvent this problem, a process called *fiber identification* is employed. First, black regions are labeled and couple of regions between which a fiber exists is selected. In the next step, the two selected regions are connected performing a *dilation* operation with a large enough *structuring element.* Dilation is an operation that grows or thickens objects in a binary image by adding pixels to the boundaries of objects. The specific manner and extent of this thickening is controlled by the size and shape of the structuring element used. In the following process, an *erosion* operation with the same structuring element is performed and the fiber is recognized. Erosion shrinks or thins objects in a binary image by removing pixels on object boundaries. As in dilation, the manner and extent of shrinking is controlled by a structuring element. Then, in order to enhance the processing speed, the image is cropped to the size of selected regions. Afterwards, fiber diameter is measured according to the previously explained algorithm. This trend is continued until all of the fibers are analyzed. Finally, the data in pixels may be converted to *nm* and the histogram of fiber diameter distribution is plotted. Figure 13. shows a labeled simple simulated image and the histogram of fiber diameter obtained by this method.

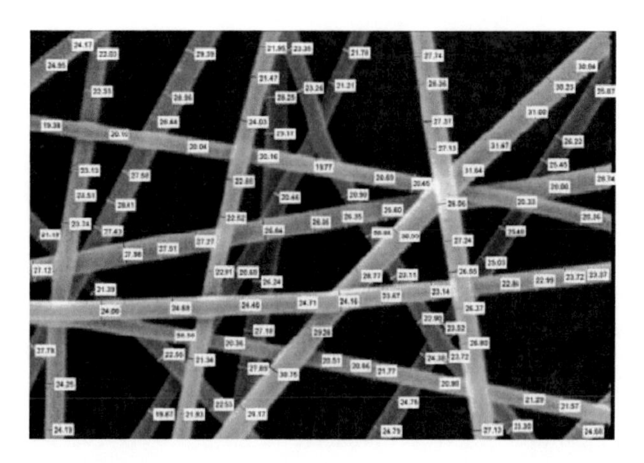

Figure 10. Manual method.

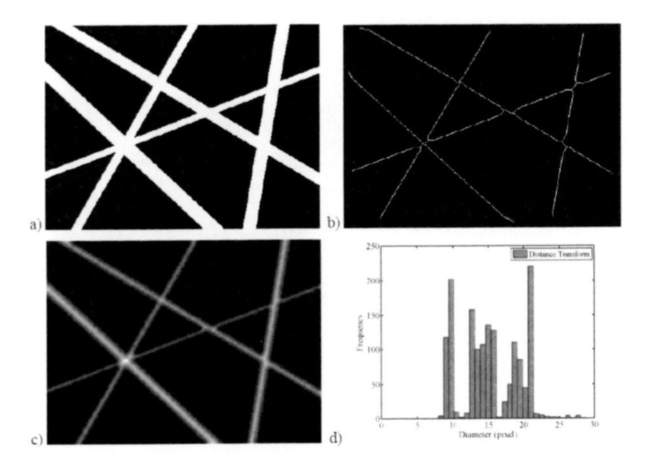

Figure11. a) A simple simulated image, b) Skeleton of (a), c) Distance map of (a) after pruning, d) Histogram of fiber diameter distribution obtained by distance transform method.

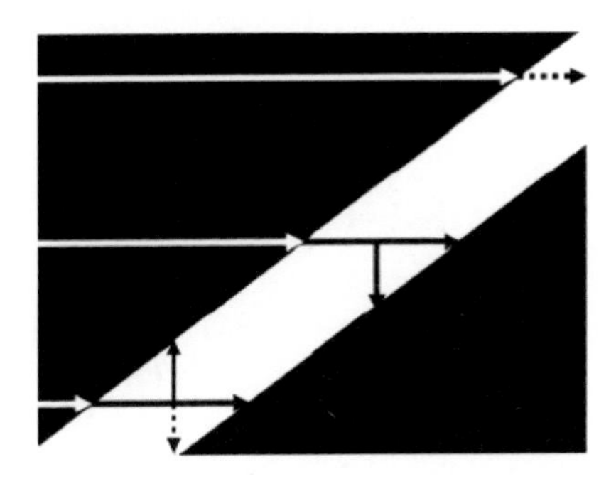

Figure 12. Diameter measurement based on two scans in direct tracking method.

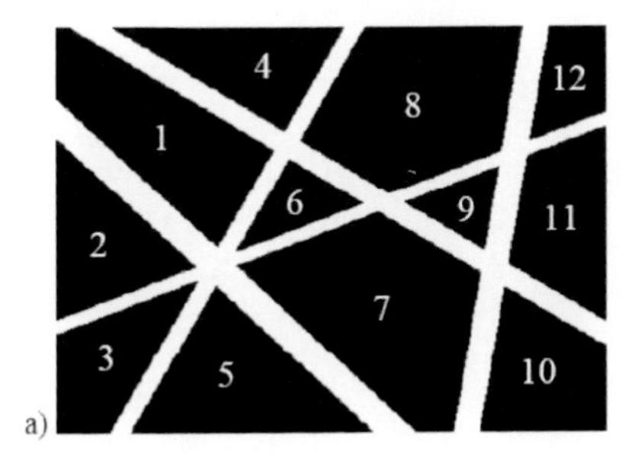

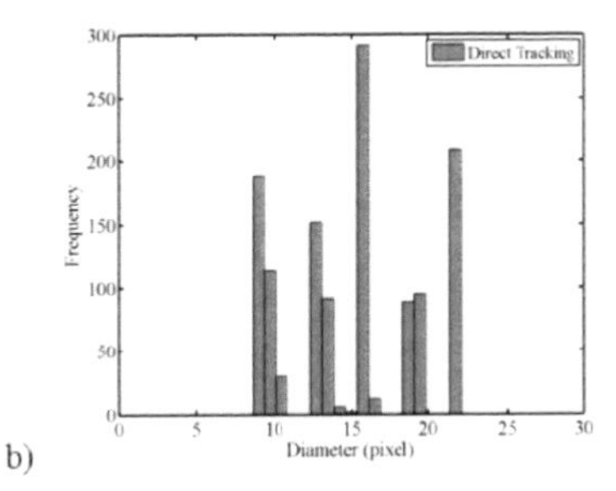

Figure 13. a) A simple simulated image which is labeled, b) Histogram of fiber diameter distribution obtained by direct tracking.

Conclusion

In this study, effect of laminating temperature on nanofiber/laminate properties investigated to make next-generation protective clothing. First, surface images of nanofiber web after lamination were taken using optical microscope in order to consider morphology changes. It was observed that nanofiber web remains unchanged as laminating temperature is below PPSN melting point. In addition, to compare breathability of laminates, air permeability was measured. It was found that by increasing laminating temperature, air permeability was decreased. Furthermore, it only was observed that the adhesive force between layers in laminate was increased with temperature rise. The mechanical properties of the samples laminated by electrospun nanofibers showed significant impeovements.

These results indicated that laminating temperature is effective parameter for lamination of nanofiber web into fabric structure. Thus, by varying this parameter could developed fabrics with different levels of thermal comfort and protection depending on our need and use.

References

[1] H. Fong and D.H. Reneker, *Electrospinning and the formation of nanofibers*, in: D.R. Salem (Ed.), *Structure formation in polymeric fibers*, Hanser, Cincinnati (2001).

[2] D. Li and Y. Xia, *Adv. Mater.*, **16**, 1151 (2004).

[3] R. Derch, A. Greiner and J.H. Wendorff, *Polymer nanofibers prepared by electrospinning*, in: J.A. Schwarz, C.I. Contescu and K. Putyera (Eds.), *Dekker encyclopedia of nanoscience and nanotechnology*, CRC, New York (2004).

[4] M. Ziabari, V. Mottaghitalab, A. K. Haghi, Simulated image of electrospun nonwoven web of PVA and corresponding nanofiber diameter distribution, *Korean Journal of Chemical engng* , Vol.25, No. 4, pp. 919-922 ,2008.

[5] M. Ziabari, V. Mottaghitalab, A. K. Haghi, Evaluation of electrospun nanofiber pore structure parameters, *Korean Journal of Chemical engng* , Vol.25, No. 4, pp. 923-932 ,2008.

[6] M. Ziabari, V. Mottaghitalab, A. K. Haghi, Distance transform algoritm for measuring nanofiber diameter, *Korean Journal of Chemical Eng*ng, Vol 25, No. 4, pp. 905-918, 2008.

In: Electrospinning Process and Nanofiber... ISBN 978-1-61209-330-7
Editors: A.K. Haghi and G.E. Zaikov

Chapter 7

SIGNIFICANT IMPROVEMENTS IN PRODUCTION AND SIMULATION OF ELECTROSPUN FIBERS CHITOSAN/CARBON NANOTUBE

A.K. Haghi [*]
University of Guilan, Rasht, Iran

ABSTRACT

In this research for the first time the chitosan/carbon nanotube nanocomposite fabric have been successfully prepared by electrospinning and is reported as a significant improvement. The acetic acid (1-100%), formic acid, and trifluoroacetic acid/dichloromethane (TFA/DCM) mixture (the volume ratio of 70:30) was tested as solvent. The electrospun nonwoven fabric was characterized by Scanning electronic microscopic (SEM) imaging. Under optimization conditions, homogenous chitosan/carbon nanotube nanofibers with a mean diameter of 455 nm and known physical characteristics were prepared. In the second part, a simulation algorithm has been applied for generating nonwovens with known characteristics. Since the physical characteristics of simulated images are known exactly, one can employ them to test the

---

* Haghi@Guilan.ac.ir

usefulness of algorithm for characterizing diameter and other structural features.

Keywords: carbon nanotube, chitosan, electrospinning, biocomposite, nanocomposite, Industrial tool.

1. Introduction

Over the recent decades, fabrication of polymer nanofibers used in many biomedical applications such as tissue engineering, drug delivery, wound dressing, enzyme immobilization etc [1].

The nanofiber fabrications have unique characteristic such as very large surface area, ease of functionalisation for various purposes and superior mechanical properties. The electrospinning with simple process is an important technique which can be used for the production of polymer nanofibers with diameter in the range from several micrometers down to ten of nanometers.

In electrospinning, the charged jets of a polymer solution which are collected on a target are created by using an electrostatic force. Many parameters can influence in quality of fibers including the solution properties (polymer concentration, solvent volatility and solution conductivity), governing variables (flow rate, voltage, and distance between tip-to-collector), and ambient parameters (humidity, solution temperature, and air velocity in the electrospinning chamber) [2].

On the recent years, scientists interested to the electrospun of natural materials such as collagen [3,4], fibrogen [5], gelatin [6], silk [7], chitin [8] and chitosan [9,10] because of high biocompatible and biodegradable properties. Chitin is the second abundant natural polymer in the world and Chitosan (poly-(1-4)-2-amino-2-deoxy-β-D-glucose) is the deacetylated product of chitin [11].

Researchers interested to this natural polymer because of properties, including its solid-state structure and the chain conformations in the dissolved state [12]. First time, the electrospinning of pure chitosan was done by Ohkawa et al. in 2004. They prepared homogenous chitosan fibers with a mean diameter of 330nm by used TFA/DCM 70:30 [wt/wt] co-solvent in 5wt.% chitosan[13].

To improving the properties of electrospun fabrics, the synthetic materials blended to natural materials because of the lack the desired mechanical property of natural materials [14]. Carbon nanotube is one of the important synthetic polymers that were discovered first by Iijima in 1991 [15]. The scientists are becoming more interesting to carbon nanotube for existence of unique conductivity [16], mechanical, and thermal properties in this promising materials for feature applications [17]. We can produce a composite with more interfacial adhesion between the carbon nanotubes and polymers by using of the carbon nanotube as reinforcement to a polymer, and therefore enhance mechanical properties in composite because of high aspect ratio and young modulus of carbon nanotube [18,19]. The chitosan/carbon nanotube composite can be produced by the hydrogen bonds due to hydrophilic positively charged polycation of chitosan due to amino groups and hydrophobic negatively charged of carbon nanotube due to carboxyl, and hydroxyl groups [20].

The present paper discusses electrospinning of chitosan/carbon nanotube dispersion. At the first, chitosan/carbon nanotube dipersion prepared. This dispersion was electrospun and the chitosan/carbon nanotube electrospun fabric was optimized. The SEM images show homogenous chitosan/carbon nanotube nanofiber with a mean diameter of 455 nm.

2. Experimental

2.1. Materials

Chitosan polymer (degree of deacetylation of 85% and molecular weight of $5x10^5$) supplied by Sigma-Aldrich, the multi walled carbon nanotube used in this study, supplied by Nutrino, have an average diameter of 4 nm, purity of about 98%. All the reagents were commercially available and in analytical grade. The chemicals were all used as received without further purification.

2.2. Electrospinning of Chitosan/Carbon Nanotube Dispersion

Multi walled carbon nanotube was sonicated for 10 min in solvent and then stirred for 24 hr. A bout 3 ml of chitosan/carbon nanotube dispersion was placed into a 5 ml syringe with a stainless steel needle having an inert diameter of 0.6 mm was connected to positive electrode. An aluminum foil, used as the collector screen was connected to the ground. A high voltage power supply Gamma High Voltage Researcher ES30P-5W was generated DC voltages in the range of 1-25 kV. The voltage and tip-to-collector distance were fixed at

18-24 kV and 4-10 cm, respectively. The electrospinning experiments were performed at room temperature.

2.3. Measurements and Characterizations

Scanning electron microgragbs (SEM) were obtained with a philips XL30 at an acceleration voltage of 20 kV.

3. Results and Discussion

The different solvents including acetic acid 1-90%, formic acid, and TFA/DCM tested for the electrospinning of chitosan/carbon nanotube. No jet was seen upon applying the high voltage even above 25 kV by using of acetic acid 1-30% and formic acid as the solvent for chitosan/carbon nanotube. When the acetic acid 30-90%, used as the solvent, beads were deposited on the collector. Therefore, under these conditions, an electrospun fiber of carbon nanotube/chitosan could not be obtained (data not shown).

Figure 1 shows Scanning electronic microgragbs of the carbon chitosan/nanotube electrospun fibers in different concentration of chitosan in TFA/DCM (70:30) solvent. As presented in figure 1.a, at low concentrations of chitosan the beads deposited on the collector and thin fibers coexited among the beads. When the concentration of chitosan was increased as shown in figure 1.a-c the beads was decreased. Figure 1.c. show homogenous electrospun fibers with minimum beads, thin fibers and interconnected fibers.

More increasing of concentration of chitosan lead to increasing interconnected fibers at figure 1.d-e. The average diameter of chitosan/carbon nanotube electrospun fibers were increased by increasing concentration of chitosan in the Scanning electronic microgragbs 1 a-e.

Hence, chitosan/carbon nanotube solution in TFA/DCM (70:30) with 10 wt% chitosan resulted as optimization conditions for electrospinning of this solution with an average diameter of 455 nm (figure 1.c: diameter distribution, 306-672).

To understanding effects of voltage on morphologies of chitosan/carbon nanotube electrospun fibers, the SEM images at figure 2 were analyzed. When the voltage was low the beads and some little fiber deposited on collector (figure 2.a). As shown in figure 2.a-d, the beads decreased by increasing voltage from 18 kV to 24 kV for electrospinning of fibers. The average diameter of fibers prepared by 18 kV measured 307 nm. When the applied voltage increases, the average fiber diameters and their distributions increase. The average diameter of fibers for 20 kV (2b), 22 kV (2c), and 24 kV(2d), respectively, was 308 (194-792), 448 (267-656), 455 (306-672).

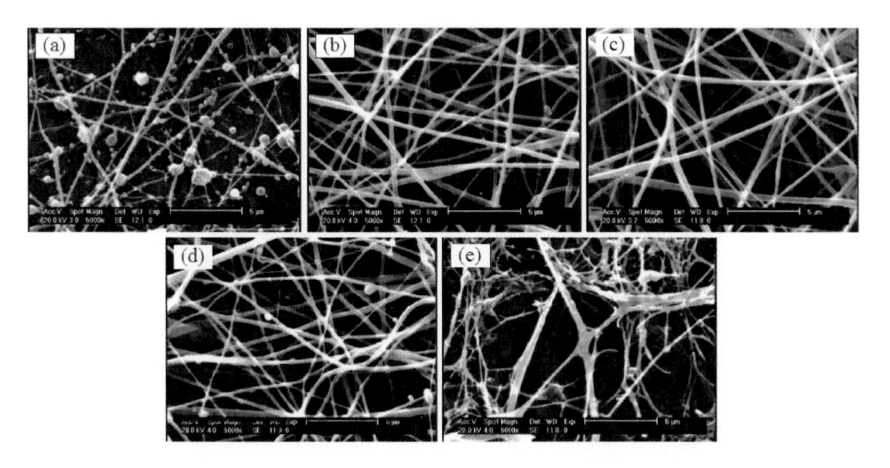

Figure 1. Scanning electron microgragh s of electrospun fibers at different chitosan concentration (wt%): (a) 8, (b) 9, (c) 10, (d) 11, (e) 12, 24 kV, 5 cm, TFA/DCM: 70/30.

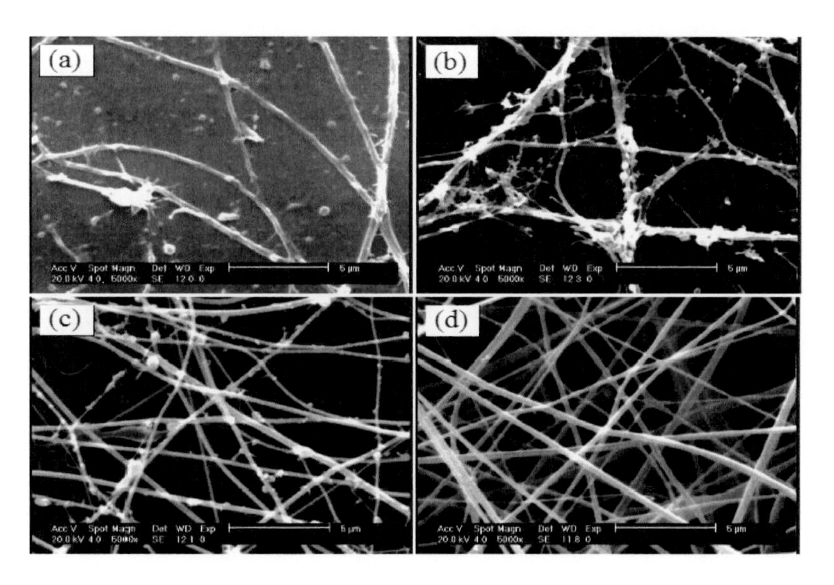

Figure 2. Scanning electronic microgragh s of electrospun fibers at different voltage (kV): (a) 18, (b) 20, (c) 22, (d) 24, 5 cm, 10 wt%, TFA/DCM: 70/30.

The morphologies of chitosan/carbon nanotube electrospun fibers at different distance tip-to-collector in figure 3, represented. When the distance tip-to-collector is low the solvent could not vapor, hence, a little interconnected fiber with high fiber diameter deposited on the collector (figure 3a). In 5 cm distance tip-to-collector (figure 3.b) obtained rather homogenous fibers with negligible beads and interconnected fibers. However, the beads increased by increasing of distance tip-to-collector as represented from figure 3b to figure 3.f. Also, the results show that the diameter of electrospun fibers decreased by increasing of distance tip-to-collector in figure 3.b, 3.c, 3.d, respectively, 455 (306-672), 134 (87-163), 107 (71-196). The fibers prepared with 8 cm (figure 3.e) and 10 cm (figure 3.f), the defects and non homogenous diameter fiber was remarkable. However, 5 cm for distance tip-to-collector was seen proper for electrospinning.

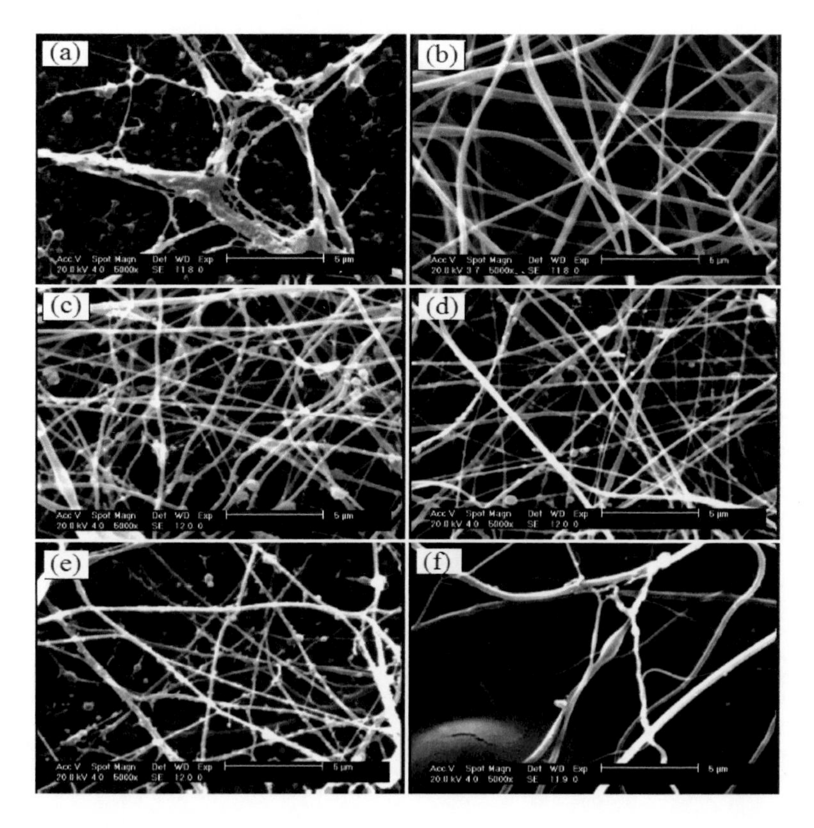

Figure 3. Scanning electronic micrograghs of electrospun fibers of Chitosan/Carbon nanotubes at different tip-to-collector distances (cm): (a) 4, (b) 5, (c) 6, (d) 7, (e) 8, (f) 10, 24 kV, 10 wt%, TFA/DCM: 70/30

4. Fiber Diameter Measurement

The first step in determining fiber diameter is to produce a high quality image of the web, called micrograph, at a suitable magnification using electron microscopy techniques. The methods for measuring electrospun fiber diameter are described in following sections.

4.1. Manual Method

The conventional method of measuring the fiber diameter of electrospun webs is to analyze the micrograph manually. The manual analysis usually consists determining the length of a pixel of the image (setting the scale), identifying the edges of the fibers in the image and counting the number of pixels between two edges of the fiber (the measurements are made perpendicular to the direction of fiber-axis), converting the number of pixels to *nm* using the scale and recording the result. Typically 100 measurements are carried out (Figure 4). However, this process is tedious and time-consuming especially for large number of samples. Furthermore, it cannot be used as on-line method for quality control since an operator is needed for performing the measurements. Thus, developing automated techniques which eliminate the use of operator and has the capability of being employed as on-line quality control is of great importance.

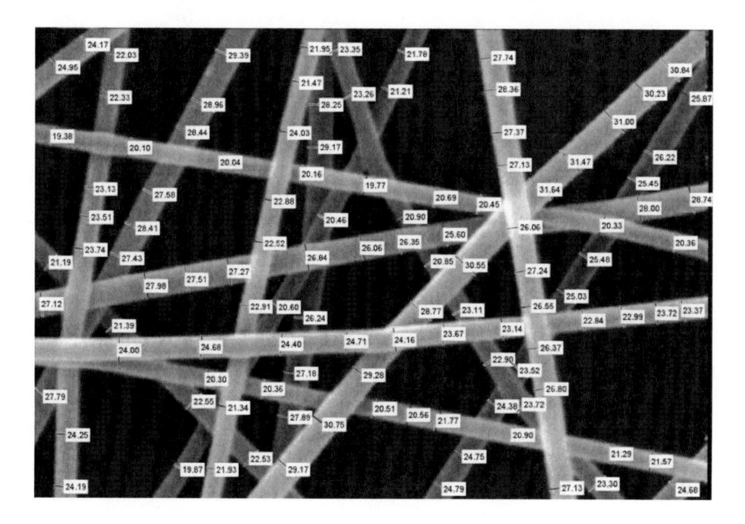

Figure 4. Manual method.

4.2. Distance Transform

The *distance transform* of a binary image is the distance from every pixel to the nearest nonzero-valued pixel. The center of an object in the distance transformed image will have the highest value and lie exactly over the object's *skeleton*. The skeleton of the object can be obtained by the process of *skeletonization* or *thinning*. The algorithm removes pixels on the boundaries of objects but does not allow objects to break apart. This reduces a thick object to its corresponding object with one pixel width. Skeletonization or thinning often produces short spurs which can be cleaned up automatically with a *pruning* procedure. The algorithm for determining fiber diameter uses a binary input image and creates its skeleton and distance transformed image. The skeleton acts as a guide for tracking the distance transformed image by recording the intensities to compute the diameter at all points along the skeleton. Figure 5. shows a simple simulated image, which consists of five fibers with diameters of 10, 13, 16, 19 and 21 pixels, together with its skeleton and distance map including the histogram of fiber diameter obtained by this method.

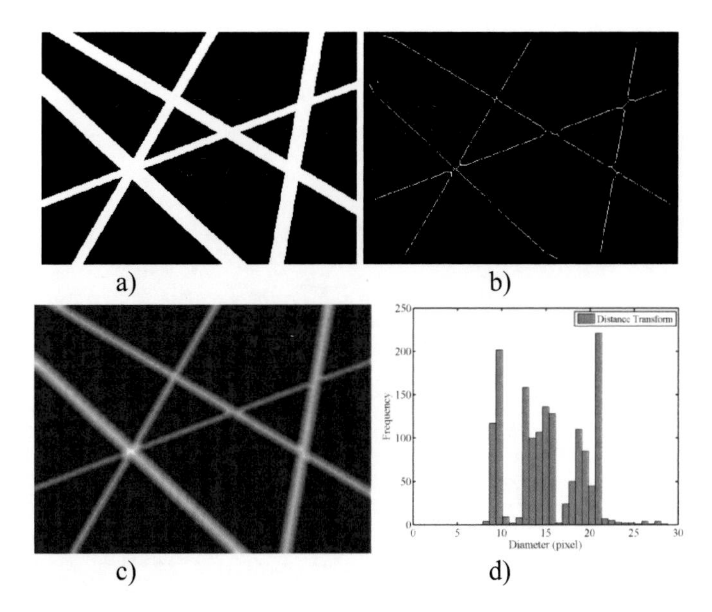

Figure 5. a) A simple simulated image, b) Skeleton of (a), c) Distance map of (a) after pruning, d) Histogram of fiber diameter distribution obtained by distance transform method.

4.3. Direct Tracking

Direct tracking method uses a binary image as an input data to determines fiber diameter based on information acquired from two scans; first a horizontal and then a vertical scan. In the horizontal scan, the algorithm searches for the first white pixel adjacent to a black. Pixels are counted until reaching the first black.

The second scan is then started from the mid point of horizontal scan and pixels are counted until the first black is encountered. Direction changes if the black pixel isn't found.

Having the number of horizontal and vertical scans, the number of pixels in perpendicular direction which is the fiber diameter could be measured from a geometrical relationship. The explained process is illustrated in Figure 6.

In electrospun nonwoven webs, nanofibers cross each other at intersection points and this brings about the possibility for some untrue measurements of fiber diameter in these regions. To circumvent this problem, a process called *fiber identification* is employed.

First, black regions are labeled and couple of regions between which a fiber exists is selected. In the next step, the two selected regions are connected performing a *dilation* operation with a large enough *structuring element.* Dilation is an operation that grows or thickens objects in a binary image by adding pixels to the boundaries of objects.

The specific manner and extent of this thickening is controlled by the size and shape of the structuring element used [21].

In the following process, an *erosion* operation with the same structuring element is performed and the fiber is recognized. Erosion shrinks or thins objects in a binary image by removing pixels on object boundaries.

As in dilation, the manner and extent of shrinking is controlled by a structuring element. Then, in order to enhance the processing speed, the image is cropped to the size of selected regions. Afterwards, fiber diameter is measured according to the previously explained algorithm. This trend is continued until all of the fibers are analyzed.

Finally, the data in pixels may be converted to *nm* and the histogram of fiber diameter distribution is plotted.

Figure 7 shows a labeled simple simulated image and the histogram of fiber diameter obtained by this method.

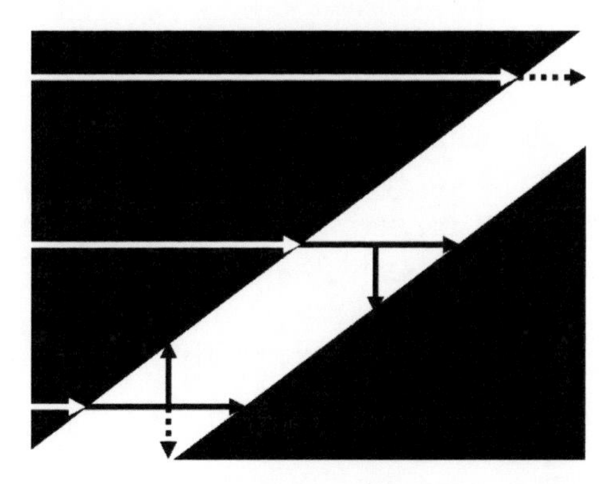

Figure 6. Diameter measurement based on two scans in direct tracking method.

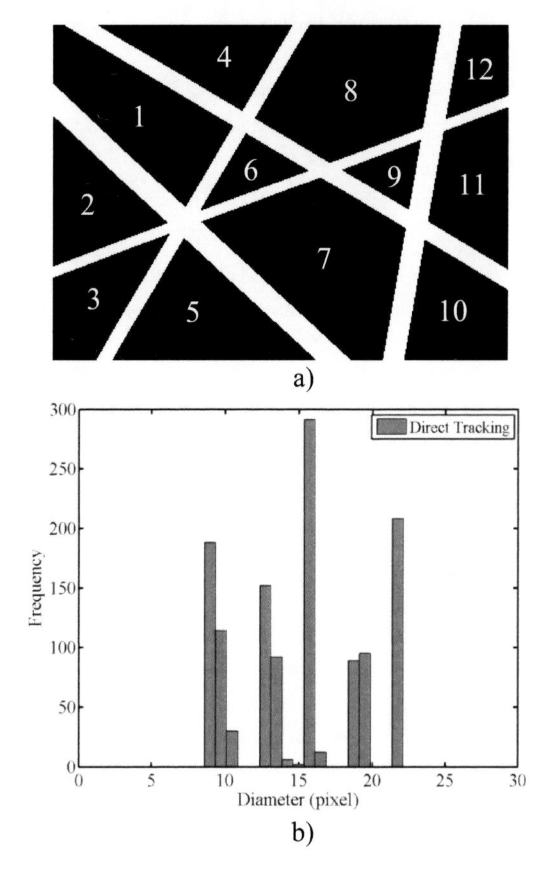

Figure 7. a) A simple simulated image which is labeled, b) Histogram of fiber diameter distribution obtained by direct tracking.

4.4. Real Webs Treatment

Both of distance transform and direct tracking algorithms for measuring fiber diameter require binary image as input.

Hence, the micrographs first have to be converted to black and white. This can be carried out by *thresholding* (known also as *segmentation*) which produces binary image from a grayscale (intensity) image. This is a critical step because the segmentation affects the result.

Prior to the segmentation, an *intensity adjustment* operation and a two dimensional *median* filter were applied in order to enhance the contrast of the image and remove noise.

In the simplest thresholding technique, called *global thresholding*, the image is partitioned using a single constant threshold.

One simple way to choose a threshold is by trial and error. Then each pixel is labeled as object or background depending on whether the gray level of that pixel is greater or less than the value of threshold respectively.

The main problem of global thresholding is its possible failure in the presence of non-uniform illumination or local gray level unevenness. An alternative to circumvent this problem is to use *local thresholding* instead. In this approach, the original image is divided to subimages and different thresholds are used for segmentation.

Another variant of this approach, which has been used in this contribution, consists of estimating the background illumination using *morphological opening* operation, subtracting the obtained background from the original image and applying a global thresholding to produce the binary version of the image.

The morphological opening is a sequential application of an erosion operation followed by a dilation operation (i.e., opening = erosion + dilation) using the same structuring element.

In order to automatically compute the appropriate threshold, *Otsu's method* could be employed [21]. This method chooses the threshold to maximize the interclass variance and minimize the intraclass variance of the black and white pixels.

As it is shown in Figure 8, global thresholding resulted in some broken fiber segments. This problem was solved in detail using local thresholding [21].

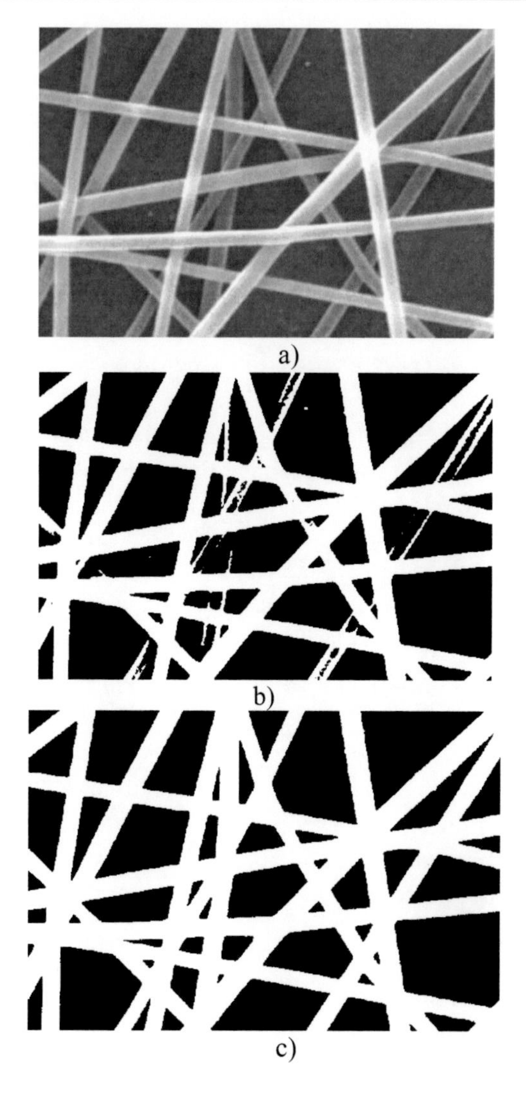

Figure 8. a) A real web, b) Global thresholding, c) Local thresholding.

5. New Distance Transform Method

The problem of the distance transform method is that skeletons are often broken at intersection points. Furthermore, since two or more fibers cross each other at the intersections, the value of the center of the object in the distance transformed image doesn't coincide with the fiber diameter because it isn't corresponding to a single fiber. As it is depicted in Figure 9.a, the intersections in the distance map are brighter than where a single fibers present. This demonstrates that higher values than expected were returned at these points. Figure 9.b. shows the broken skeleton at intersections. This problem becomes more pronounced as fibers get thicker and for points where more fibers cross each other. Hence, the distance transform method fails in measuring fiber diameter at intersections. We modified the distance transform method so that the problems associated with the intersections are solved. Furthermore, in the method proposed in [21], city block distance transform was used which as mentioned earlier, is not a realistic metric since it does not preserve the isotropy. In order to provide more rational results, in this approach we used Euclidean distance metric. The method uses a binary image as an input. Then, the distance transformed image and its skeleton are created. In order to solve the problem of the intersections, these points are identified and deleted from the skeleton.

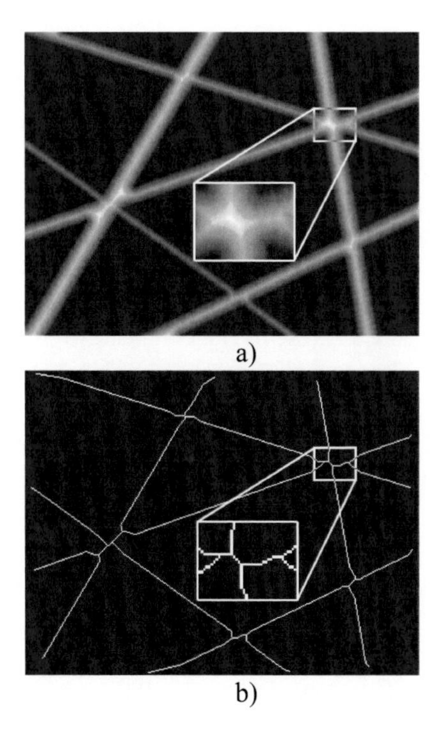

Figure 9. Distance transform method failure at intersection points: a) distance map of the image shown in Figure 8.a, b) Broken skeleton obtained from thinning of Figure 8.a. (area around an intersection has been magnified for more clarity).

First, in order to find the intersection points, a *sliding neighborhood operation* is employed. A sliding neighborhood operation is an operation which is applied to a pixel at a time; the value of that pixel in the output image is determined by the implementation of a given function to the values of the corresponding input pixel's neighborhood (Figure 10).

A neighborhood about a pixel, which is usually called the center point, is a square or rectangular region centered at that pixel. The operation consists of five steps:

1. Defining a center point and a neighborhood block.
2. Starting from the first (normally top left) pixel in the image.
3. Performing an operation (a function given) that involves only the pixels in the defined block.
4. Finding the pixel in the output corresponding to the center pixel in the block and setting the result of the operation as the response at that pixel.
5. Repeating steps 3 to 4 for each pixel in the input image.

Since at an intersection point, two or more fibers meet each other, it could be defined as a location where a white pixel in the skeleton has more than two neighboring pixels each leading a branch. Hence, performing a sliding neighborhood operation on the skeleton with a 3-by-3 sliding block and summation as the function (which is applied over all pixels in the block), the intersections could be identified as the points having values more than 3. This is demonstrated in Figure 11. (the intersections are shown with arrows). After the intersection points are located, the next step is to find the width of each one. This is carried out using the distance map of the binary input image via finding the pixel corresponding to that intersection point. The value of the distance map at the pixel is then considered as the width of that intersection.

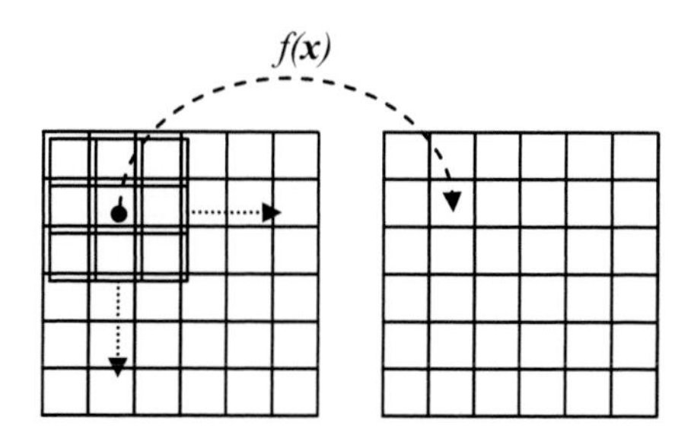

Figure 10. Sliding neighborhood operation with a 3-by-3 neighborhood block.

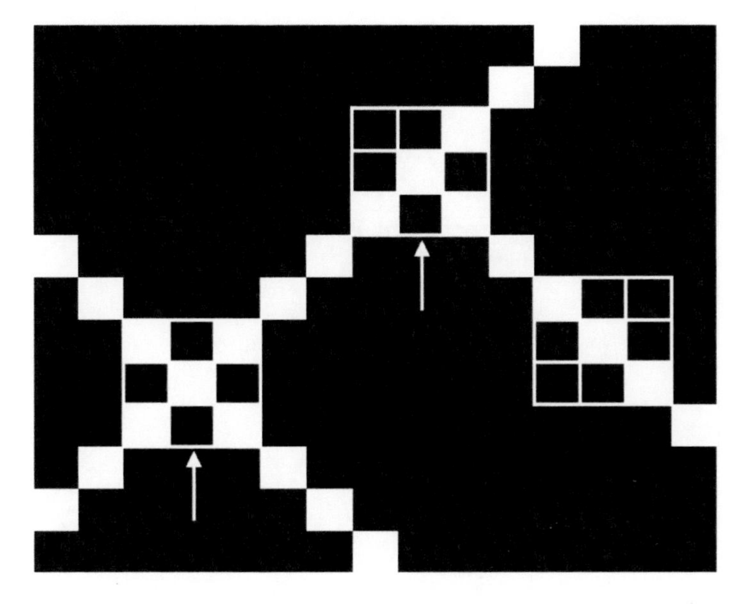

Figure 11. Identifying intersection points using a sliding neighborhood operation with a 3-by-3 neighborhood block.

After that, the pixels in the skeleton which lie inside a square with that width around the intersection point are cleaned. This procedure is replicated until each intersection is identified and cleaned. Figure 12.a. exhibits the skeleton of the simple simulated image shown in Figure 6.a. after deleting the intersection points followed by a pruning procedure.

Finally, the resultant skeleton (of which the intersections are deleted) is used as a guide for tracking the distance transformed image and fiber diameters are obtained by recording the intensities to at all points along the skeleton (white pixels in Figure 10.a. show the skeleton) and doubling the results.

The distance map of image in Figure 6.a. is also shown in Figure 12.b. for better understanding of the procedure. Setting the length of a pixel in the image, the values may then be converted to *nm* and the histogram of fiber diameter distribution is plotted. Figure 12.c. demonstrates the histogram of fiber diameter (in term of pixel) obtained by this method.

The procedure for determining fiber diameter via this approach is summarized in Figure 13. The method is efficient, reliable, accurate and so fast and has the capability of being used as an on-line method for quality control.

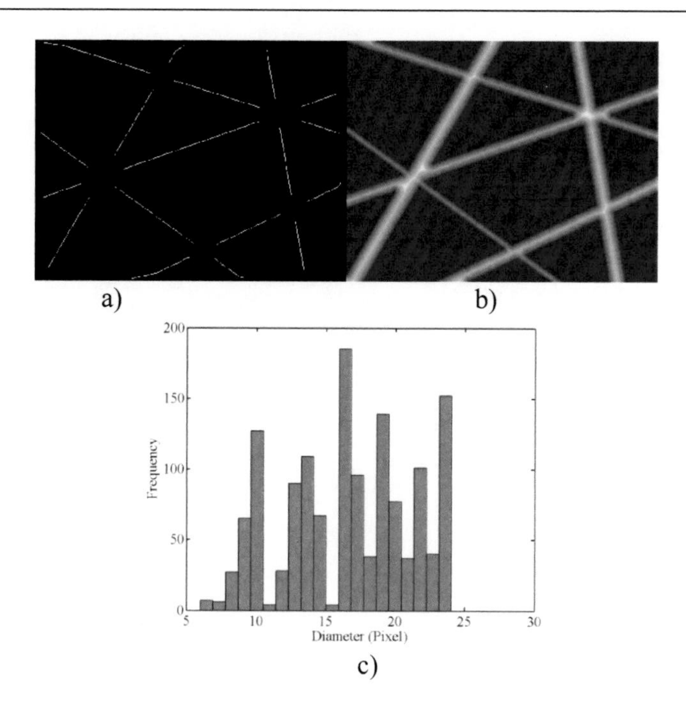

Figure 12. New distance transform method: a) the skeleton of the simple simulated image shown previously after deleting the intersection points, b) the distance map, c) histogram of fiber diameter distribution.

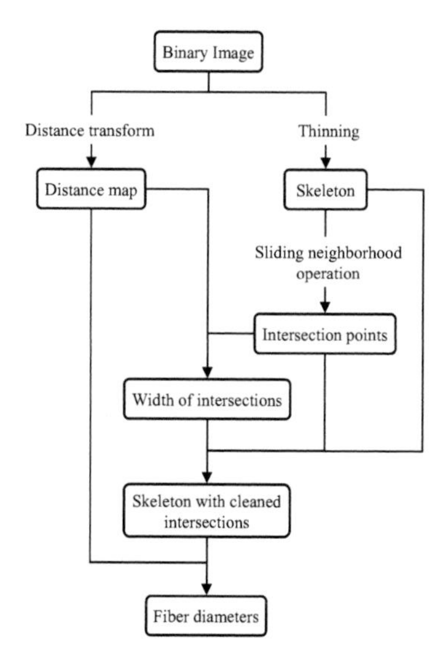

Figure 13. Flowchart of the new distance transform method.

CONCLUSIONS

Several solvent including acetic acid 1-90%, formic acid, and TFA/DCM (70:30) were investigated in the electrospinning of chitosan/carbon nanotube dispersion. It is observed that the TFA/DCM (70:30) solvent was only solvent that resulted for electrospinnability of chitosan/carbon nanotube. This is a significant improvement in electrospinning of chitosan/carbon nanotube dispersion. It is also observed that the homogenous fibers with an average diameter of 455 nm (306-672) could be prepared with chitosan/carbon nanotube dispersion in TFA/DCM 70:30. Meanwhile, the SEM images show that the fiber diameter decreased by decreasing of voltage and increasing distance tip-to-collector. A simulation algorithm was also presented for generated nonwovens with known characteristics. This could be used as an important industrial tool in characterizing diameter and other structural features.

REFERENCES

[1] Seema Agarwal, Joachim H. Wendorff, Andreas Greiner, Use of electrospinning technique for biomedical applications, *Polymer* 49 (2008) 5603–5621.

[2] Zheng-Ming Huang, Y.-Z. Zhang, M. Kotaki, S. Ramakrishna, A review on polymer nanofibers by electrospinning and their applications in nanocomposites, *Composites Science and Technology* 63 (2003) 2223–2253.

[3] Jamil A. Matthews, Gary E. Wnek, David G. Simpson, Gary L. Bowlin, Electrospinning of Collagen Nanofibers, *Biomacromolecules* 2002, 3, 232-238.

[4] Kyong Su Rho, Lim Jeong, Gene Lee, Byoung-Moo Seo, Yoon Jeong Park, Seong-Doo Hong, Sangho Roh, Jae Jin Cho, Won Ho Park, Byung-Moo Min, Electrospinning of collagen nanofibers: Effects on the behavior of normal human keratinocytes and early-stage wound healing, *Biomaterials* 27 (2006) 1452–1461.

[5] Michael C. McManus, Eugene D. Boland, David G. Simpson, Catherine P. Barnes, Gary L. Bowlin, Electrospun fibrinogen: Feasibility as a tissue engineering scaffold in a rat cell culture model, *InterScience*. DOI: 10.1002/jbm.a.30989.

[6] Zheng-Ming Huang, Y.Z. Zhang, S. Ramakrishna, C.T. Lim, Electrospinning and mechanical characterization of gelatin nanofibers, *Polymer* 45 (2004) 5361–5368.

[7] Xiaohui Zhang, Michaela R. Reagan, David L. Kaplan, Electrospun silk biomaterial scaffolds for regenerative medicine, *Advanced Drug Delivery Reviews* 61 (2009) 988–1006.

[8] Hyung Kil Noh, Sung Won Lee, Jin-Man Kim, Ju-Eun Oh, Kyung-Hwa Kim, Chong-Pyoung Chung, Soon-Chul Choi, Won Ho Park, Byung-Moo Min , Electrospinning of chitin nanofibers: Degradation behavior and cellular response to normal human keratinocytes and fibroblasts, *Biomaterials* 27 (2006) 3934–3944.

[9] Kousaku Ohkawa, Ken-Ichi Minato, Go Kumagai, Shinya Hayashi, and Hiroyuki Yamamoto, Chitosan Nanofiber, *Biomacromolecules 2006*, 7, 3291-3294.

[10] Xinying Geng, Oh-Hyeong Kwon, Jinho Jang, Electrospinning of chitosan dissolved in concentrated acetic acid solution, Biomaterials 26 (2005) 5427–5432.

[11] O. C. Agboh and Y. Qin, Chitin and Chitosan Fibers, Polymers for Advanced Technologies 8 (1997) 355–365. [12] R. Marguerite, Chitin and chitosan: properties and applications, *Prog. Polym. Sci.* 31 (2006) 603-632.

[12] Kousaku Ohkawa, Dongil Cha, Hakyong Kim, Ayako Nishida, Hiroyuki Yamamoto, electrospinning of chitosan, *Macromol. Rapid Commun.* 2004, 25, 1600–1605.

[13] Travis J. Sill, Horst A. von Recum, Electrospinning: Applications in drug delivery and tissue engineering, *Biomaterials* 29 (2008) 1989e2006.

[14] S. Iijima, Helical microtubules of graphitic carbon, *Nature* 354 (1991) 56–58.

[15] Nathalie K. Guimard, Natalia Gomez, Christine E. Schmidt, Conducting polymers in biomedical engineering, *Prog. Polym. Sci.* 32 (2007) 876–921.

[16] Marek Trojanowicz, Analytical applications of carbon nanotubes: a review, *Trends in Analytical Chemistry* 25 (2006) 480-489.

[17] Shao-Feng Wang, Lu Shen, Wei-De Zhang, and Yue-Jin Tong, Preparation and Mechanical Properties of Chitosan/Carbon Nanotubes Composites, *Biomacromolecules* 6 (2005) 3067-3072.

[18] Yeong-Tarng Shieh, Yu-Fong Yang, Significant improvements in mechanical property and water stability of chitosan by carbon nanotubes, *European Polymer Journal* 42 (2006) 3162–3170.
[19] Ying-Ling Liu, Wei-Hong Chen, Yu-Hsun Chang, Preparation and properties of chitosan/carbon nanotube nanocomposites using poly(styrene sulfonic acid)-modified CNTs, *Carbohydrate Polymers* 76 (2009) 232–238.
[20] Ziabari, M.; Mottaghitalab, V.; Haghi, A. K. *Korean. J. Chem. Eng.* 2008, 25, 919-922.

In: Electrospinning Process and Nanofiber... ISBN 978-1-61209-330-7
Editors: A.K. Haghi and G.E. Zaikov

Chapter 8

ELECTROSPINNING FOR TISSUE ENGINEERING APPLICATIONS

Joseph Lowery[1], Silvia Panseri[2], Carla Cunha [2,3] and Fabrizio Gelain [2,3]

[1]Institute for Soldier Nanotechnologies, Massachusetts Institute of Technology, 500 Technology Sq, Cambridge, USA
[2]Biotechnology and Biosciences Department, University of Milan - Bicocca, Piazza della Scienza 2, Milan, 20126, Italy
[3]Center for Nanomedicine and Tissue Engineering - A.O. Ospedale Niguarda Ca' Granda, Piazza dell'ospedale maggiore 3, Milan, 20162 Italy

ABSTRACT

Electrospinning is one of three techniques available nowadays for the processing of fibers mimicking the extracellular environment at the nanoscale, the so-called nanofibers. This technique allows the fabrication of a controllable continuous nanofiber scaffold made of natural polymers, of synthetic polymers or of inorganic substances. Moreover, through secondary processing, the nanofiber surface can be functionalized to display specific biochemical characteristics.

This chapter will discuss/summarize in detail the currently available electrospinning techniques, recent trends on nanofiber processing and characterization and their current biomedical applications, with particular emphasis on the most recent tissue engineering applications for regenerative medicine.

Keywords: electrospinning, nanofiber, tissue engineering.

1. Electrospinning Techniques

1.1. Electrospinning Overview

Electrospinning is one of current 5 processes to fabricate nanofibers: drawing, template synthesis, phase separation, self-assembly and electrospinning. Electrospinning is the most widely studied, since it has demonstrated the most promising results in terms of tissue engineering applications and probably is the only process with the potential for mass production [1].

The ability to generate polymer fibers from an electrically charged jet was discovered and patented in various forms at the beginning of the 20th Century [2, 3]. The first patents were developed by Cooley and Morton, and deal primarily with the dispersion of electrically charged fluids. Formhals' patents in the 1930's and 40's [4-7] along with Norton [8], who addressed specific work with polymer melts, round out the earliest work in the field.

The versatility of the electrospinning process presents a unique opportunity to synthesize tissue engineering constructs from a wide variety of materials and with a great degree of control.

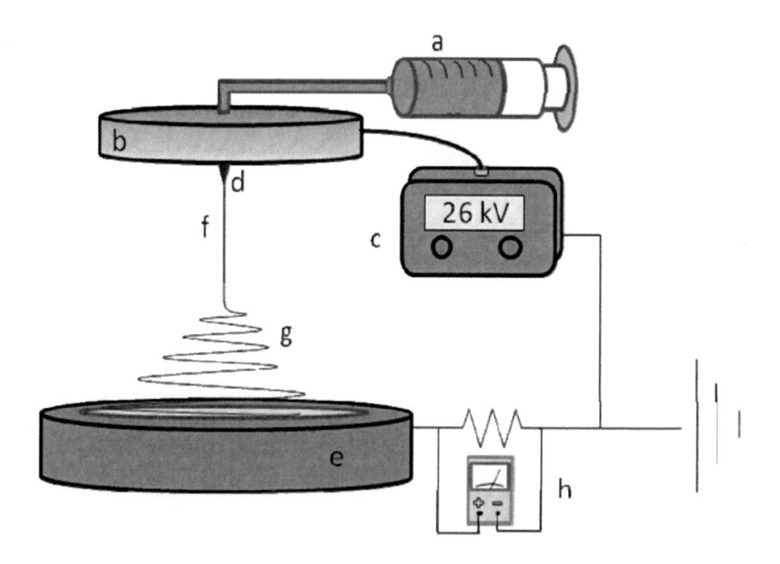

Figure 1. Basic parallel-plate electrospinning apparatus.

Figure 1 presents a basic parallel-plate electrospinning setup. Polymer solution (a) is pumped through a steel nozzle located in a plate maintained at a high electric potential (b), where the solution develops excess electric charge. Charge is generated from a DC power supply (c). At the tip of the protruding nozzle the meniscus is deformed into a cone of fluid called the Taylor cone (d), from which a single jet of solution proceeds downfield toward the grounded target (e). After a period of linear elongation (f), the jet gives way to "whipping" (g). Solvent evaporates from the jet during the whipping process and fiber is deposited on the grounded target as solid polymer. A multimeter (h) in series with a resistor measures the current conducted by the jet for purposes of monitoring jet stability, but is not a requirement to production. Fibers build up on the grounded target, depositing with random directionality. The final product is a tangible nonwoven mat with high porosity (~70%-90).

Many of the pieces in Figure 1 are capable of being altered to the user's desire. The parallel plate is not a requirement; its purpose is to provide an even electric field from the needle tip to the grounded target. A single charged needle is capable of generating electrospun fibers, and possibly at a lower voltage. The syringe pump is another luxury that many researchers do not include. A glass pipette placed at a 30° to 45° angle will induce flow of the viscous polymer solution within, though not at a uniform rate (in this case a copper wire within the glass pipette is connected to the power source to charge the fluid). In exchange for simplicity the users risks the uniformity of the electrospun product, as a reduced flow rate will produce smaller fiber diameters. Consequently, the syringe pump should be included whenever possible. Orientation of the target relative to the charged plate may also be varied. Figure 1 presents a vertical orientation; a horizontal orientation and variations in between are also possible. Both systems present unique advantages. Vertical orientation ensures a more uniform thickness of the product since gravity acts in the direction of the flow. The jet cannot flow perfectly horizontal due to the force of gravity, and the resulting fiber mat may be of an uneven thickness. Vertical orientation risks damage to the electrospun product due to a droplet of polymer solution falling on to the mat (un-electrospun droplets of polymer in solution can occur when the flow rate and voltage are not in sync, during voltage startup, or at other times). Since the needle is directly above the target, any droplets will land on the grounded target where the product is being deposited. This creates a flaw in the electrospun product and possibly introduces harmful solvents into the system (though the solvent should evaporate away). A horizontally-oriented target does not run the risk of besmirchment from solution droplets. The user is left to decide for themselves which method is more useful, though an evenly-distributed product is more valuable when continuous electrospinning can be achieved. The target itself is also extremely versatile. Since the fiber mat will conform to the shape of the collector, a range of shapes are capable of being produced. In most cases, fibers are collected on a flat target – either a solid conductor such as aluminum foil or a wire mesh. Fibers tend to accumulate preferentially on the wire mesh before bridging the gaps between wires, leading to thickness disparities. The trade-off is the ability to easily remove the product from the target. As fiber diameter decreases, products become more difficult to remove from the target collector without stretching or ripping, especially in the case of solid collectors. Wire meshes have fewer contact points with the fibers, decreasing the adhesion forces (but creating a possible thickness disparity).

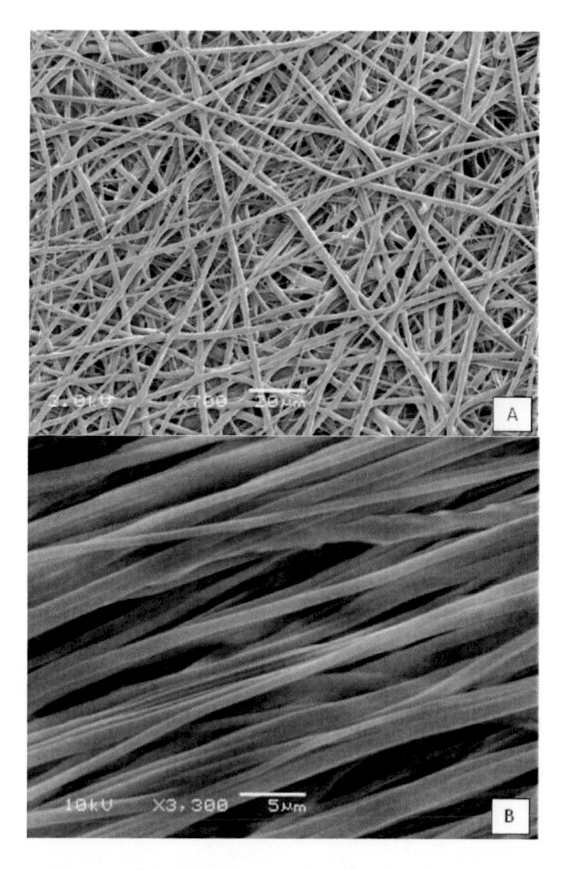

Figure 2. Scanning Electron Microscope images of electrospun fibers. *A)* As-spun randomlyoriented PCL fibers. *B)* Aligned PCL fibers.

A reduction in the distance between wires diminishes the thickness disparity (but also increases the number of contact points between the fibers and target). Alignment of the fibers along a single axis is also possible. Electrospun fibers normally deposit in a 2-D random orientation, as seen in Figure 2.A. By using a rotating mandrel target (and rotating at a speed close to that of the fiber deposition), fibers can be aligned with near-perfect orientation [9]. Two conducting targets placed a few centimeters apart will also create an area of aligned fibers in the gap between the conductors. Fibers spanning the gap search for the shortest path to the nearest conductor, leading to aligned fibers oriented perpendicular or nearly perpendicular to the conductor edge and parallel to one another (Figure 2.B).

Figure 3. Flat vs. round target electrospun products *A)* Cross-section of an electrospun fiber mat generated on a flat target. *B)* Cross-section of an electrospun fiber mat generated on a round target, creating a hollow tube morphology (Panseri et al., 2008).

Any fibers depositing solely on the conductors in this case will deposit in the regular random 2-D fashion. Lastly the shape of the conductor itself is controllable. As a most basic example, fiber mats can be shaped into hollow tubes by depositing on a cylindrical target.

Figure 3 presents electrospun mats deposited in different shapes. More complex shapes are possible, though fibers will not deposit evenly around corners without realignment of the target, nor will fibers fill depressed areas along the contour of the target for reasons listed above. Regardless, these constraints can be overcome with careful engineering to generate a wide range of material shapes.

Chemically heterogeneous mats comprised of variant fibers can also be constructed. In addition to processing a single material, there is also the possibility of electrospinning multiple components simultaneously from different solutions (resulting in a well-dispersed matrix of fibers with variant chemical structure), or creating a layered scaffold by sequentially electrospinning any number of polymer solutions [10]. Core-shell electrospinning is capable of producing fibers with an internal core material different from that of the external shell [11]. Many of these technologies have only been introduced very recently; increasingly complicated techniques, such as twin-screw electrospinning [12] may yet be developed to generate mats of specific chemical functionality.

A caution: variations in the apparatus carry the possibility of altering the resulting fiber diameter and morphology; one set of experimental parameters – solution flow rate, voltage, needle-to-target distance – may produce different results on variant equipment. These disparities could include fiber diameter, physical fiber morphology (cylindrical fibers vs. bead-on-a-string fibers), or mat thickness.

As a result, care should always be taken to include as much detail as possible when describing experimental protocol, lest reproduction of the results become difficult or even impossible.

1.2. Electrospinning Production for Tissue Engineering

In synthesizing a tissue engineering construct or scaffold, the most important design considerations are mechanical strength, degradation time, surface chemistry, and scaffold architecture. Electrospinning has the advantage of requiring a minimal amount of specialized laboratory equipment to modify three-dimensional scaffolds, therefore making it an attractive technology for the field of tissue engineering [1, 13]. The first two variables – mechanical strength and degradation time – are almost exclusively controlled by the choice of scaffold material. Fortunately, electrospinning is capable of generating polymer fibers from a wide range of materials. Commodity polymers such as polystyrene, poly(ethylene oxide), poly(methyl methacrylate), and poly(ε-caprolactone) are frequently cited in publications, but high-end materials are also frequently electrospun and published as proof-of-concept. This includes multiple types of collagen [14, 15] and even DNA. Water, chloroform, methanol, ethanol, tetrahydrofuran, dimethylformamide, hexafluoroisopropanol, and mixtures thereof are primarily used as the solvent phase, though other solvents like dimethylacetamide are not uncommon. Solvent choice is dictated by polymer solubility, boiling point, and dielectric constant. Macromolecules that are incapable of being electrospun on their own due to low molecular weight, high entanglement concentrations, or possibly cost can be electrospun either in conjunction with a second component via a blended solution (PEO is a popular choice) or core-shell electrospinning, as mentioned previously. Core-shell electrospinning generates fibers with an internal core and external shell of varying materials. By using an outer shell material that is both capable of being electrospun and degraded during post-processing (generally PEO for its solubility in water), fiber mats of the internal material can be generated. While researchers have conducted *in-vitro* cell viability tests on many materials that would serve as poor *in-vivo* implants due to their vast breakdown time or potentially toxic byproducts (for instance, polystyrene [11], polyetherimide [16], or poly(ethylene terephthalate) [17] many other materials have shown promise for *in-vivo* work. These include purchasable materials such as poly(ε-caprolactone) (PCL) [18], poly(L-lactic acid) (PLLA) [19], poly(glycolic acid) (PGA) [20], poly(lactide-co-glycolide) (PLGA) [21]. Specially-synthesized polymers [22] are also of interest. Finally, naturally occurring materials such as collagen and chitosan are a desirable choice due to their biocompatibility. These materials are often expensive or difficult to electrospun, however, and may not automatically be the best choice. Specific material choice will likely vary from one tissue-specific construct to the next, and new materials are constantly being synthesized and electrospun for this purpose.

As mentioned above, highly toxic solvents are often employed in creating polymer solutions for electrospinning. Among these is hexafluoroisopropanol (HFIP), which is commonly used for solvating materials such as collagen that are largely insoluble in other solvents. Fortunately, these solvents evaporate during the whipping portion of the electrospinning process; in the event that they do not evaporate completely, highly toxic components could potentially be released during scaffold degradation. Synthesizing a scaffold without the use of toxic solvents would clearly be more ideal, either by using water as the solvent phase or by electrospinning from the melt phase. Attempts to electrospin from the melt phase have been largely unsuccessful due to the high melt viscosity and quick cooling of the jet in-flight, though a few researchers have succeeded [23, 24]. Using an aqueous solution to electrospun materials is much more promising, but the number of water-soluble materials is limited and these materials will likely solvate during future contact with any water-based solutions such as PBS. One group has discovered a means of inhibiting re-solvation by affecting the molecular organization of the material – silk – following the electrospinning process [25]. This strategy may be unique to silk, but offers hope that materials capable of being electrospun from an inert solvent such as water can still be candidates for tissue engineering constructs.

Surface chemistry is another vital component to consider in tissue engineering constructs. As-spun mats are obviously limited to the material of construction, though it is possible to affect the surface chemistry through careful synthesis procedures and post-processing to graft on RGD peptides [26].

The final design variable – scaffold architecture – can be divided into three different categories: fiber diameter, pore diameter, and porosity. There are currently no effective means of independently varying these three critical variables. Of these variables, fiber diameter and porosity are the easiest to measure. Average fiber diameter is determined via Scanning Electron Microscope (SEM) measurements, and is readily controllable through processing variables such as flow rate, polymer concentration, solvent choice, applied voltage, and needle-to-target distance. Researchers have successfully developed a scaling argument to calculate the terminal diameter, or the minimum fiber diameter capable of being achieved under a certain set of fluid and electric field conditions [27]. Porosity can be measured by multiple techniques, including simple gravimetric measurements of length, width, height, and mass. Efforts to control porosity have mainly focused on increasing the overall porosity through solvent choice [10], and adding dispersing agents to the fiber mat during deposition to control the distance between fiber layers [28, 29]. Pore diameter measurements are much more complicated, but can be conducted by mercury porosimetry or liquid extrusion porosimetry [30, 31]. Reporting pore diameters can be equally complicated, as measurements are often reported in multiple statistics. Three examples are median pore diameter by volume, the median pore diameter by area, and the average pore diameter. All three values have very different physical significance. The median pore diameter by volume represents the pore diameter at which half the available volume is filled with mercury, easily determined from a plot of cumulative intruded volume, V, vs. pore diameter, D. The median pore diameter by area represents the median diameter for cumulative surface area, A, as a function of pore diameter. The area is calculated by dividing the differential volume intrusion at a given pore diameter by D/4, since the volume, diameter, and outer wall area of a cylinder are connected by the relationship $D = 4V/A$. While the volume of a cylinder scales as the square of the diameter, the area of the outer walls of a cylinder scales only by diameter to the first power. The median pore diameter by volume is therefore weighted to larger pores than the median pore diameter by area. Finally, the average pore diameter is calculated as 4 times the total intrusion volume V divided by the cumulative surface area, per the relationship described earlier. These three values often vary by a factor of 10 or more and do not completely address the statistical information most useful in characterizing the electrospun mat. Despite this, pore diameter is of critical importance to the design of scaffold architecture as it is the only measurement that specifically addresses quantized units of void volume available for tissue growth. Due to the concise control over porosity and pore diameter that are required to successfully build tissue engineering scaffolds via the electrospinning process, future research will likely address these two areas of interest.

Though electrospinning creates randomly interconnected void spaces throughout a scaffold, there is currently little to no way of creating a uniform, controlled, three-dimensional pore structure that can be incorporated into an electrospun scaffold. Currently, few techniques are capable of drastically affecting the scaffold porosity in a homogeneous manner. The addition of dissolvable spacers to the electrospun mat during deposition is capable of increasing the scaffold porosity; however, if the size scale of the spacer is much greater than that of the fiber diameter, inhomogeneities in the scaffold structure occur [32]. It has been found that the use of ice crystals generates the proper space between depositing fibers and can therefore greatly increase the scaffold porosity over traditionally-spun mats [28]. Another area of interest is the reduction of scaffold fiber diameters. The diameters of electrospun fibers generally reside on the upper limits of the natural ECM's 50–500 nm range. Fibers diameters even end up as micron scale, depending on the material and electrospinning solution concentration.

Despite the simplicity, diversity, and control offered by electrospinning, it is by no means a perfect solution to the creation of nanofibrous ECM analogues. Like other technologies, trade-offs are often made with regard to mechanical strength of degradation time, or scaffold porosity and fiber diameter; however, the possibility of generating increasingly biomimetic scaffolds drives the field for the introduction of new electrospinning technologies.

2. Applications of Electrospun Nanofibers in Biomedical Applications

Electrospun fibers are generally collected as two-dimensional membranes with randomly arranged structures, and this has greatly limited their application. In fact, in order to make use of the electrospinning technique in biomedical applications, it is important to fabricate fibers with controllable 3D macro and microstructures. Such a nano/microfiber scaffold presents a high surface to volume ratio and porosity and has the potential to provide enhanced cell adhesion and, due to the similarity of their 3D structure to natural ECM, they supply a micro/nano environment for cells to grow and carry out their biological functions [33, 34]. In fact, cells present typically a diameter in the range of 6-20 μm and respond to stimuli from the macro environment down to the molecular level.

Hence, nanofibrous structures have been strongly pursued as scaffolds for tissue engineering applications and for a broad range of biomedical applications.

The natural ECM is a complex structure, it consists mainly of two classes of macromolecules: nanometer diameter fibrils (collagens) and polysaccharide chains of proteoglycans and glycosaminoglycan, and it may contain other important substances such as various minerals. Embedded fibrous collagens are organized in a 3D fiber network, which provides structural and mechanical stability. The ECM of natural tissue is characterized by fiber networks with wide range of pore diameter distribution, high porosity, and effective mechanical properties. High porosity provides more structural space for cell accommodation and makes the exchange of nutrient and metabolic waste between a scaffold and environment more efficient. The fulfilment of all of these characteristics is fundamental criteria for the design of successful tissue-engineered scaffolds. The task to reproduce ECM is in fact challenging, since it meets the specific requirements of the tissue and organ in question. For instance, fibrils that compose the ECM of tendon are parallel and aligned, while those found on the skin are mesh-like.

Electrospinning not only is able to fabricate nanofibers, but moreover it is a technique that can use a wide range of materials to be electrospun, as we have seen before.

2.1. Wound Dressing

Wounds presenting large amounts of cell loss require immediate coverage with a dressing, primarily to protect the wound. An ideal dressing should mimic the functions of native skin, protecting the injury from loss of fluid and proteins, enabling the removal of exudates, inhibiting exogenous micro-organism invasion and improving aesthetic appearance of the wound site [35, 36].

Post-surgical adhesion is the most important challenge that affects wound healing and occurs with the use of either conventional bandages or barrier devices. In order to prevent post-surgical adhesion, a study used PLAGA electrospun non-woven bioabsorbable nanofiber matrices as bandages in a rat model and showed excellent anti-adhesion effect and prevented complete surgical adhesions [37].

2.2. Controlled Drug Delivery

Polymeric drug delivery systems have numerous advantages compared to conventional dosage forms, such as improving therapeutic effects reducing toxicity, convenience, etc. Pharmaceutical Release dosage can be designed as rapid, immediate, delayed, pulsed, or modified dissolution depending on the polymer carriers used and other included additives.

Biodegradable polymers have been made into fibrous scaffolds as drug carriers. The main advantage of fibrous carriers is that they offer site-specific delivery of any number of drugs from the scaffold into the body. In addition, the drug can be capsulated directly into fibers with different sizes, and these systems have special properties and surprising results for drug release different from other formulations.

Drug delivery with polymer nanofibers is based on the principle that the dissolution rate of a drug particulate increases with increased surface area of both the drug and the corresponding carrier if necessary. Furthermore, unlike common encapsulation involving some complicated preparation process, therapeutic compounds can be conveniently incorporated into the carrier polymers using electrospinning.

The resulting nanofibrous membrane containing drugs can be applied topically for skin and wound healing or post-processed for other kinds of drug release. Thus, electrospinning show potential as an alternative polymer fabrication technique to drug release systems from particles to fibers.

3. Applications of Electrospun Nanofibers in Tissue Engineering

3.1. Historical and Definition

Tissue regeneration has been relying mainly on autologous (within the same individual) or allogenous (between different individuals) cell or tissue transplantation.

However, the former has limitations such as donor site morbidity and limited availability and the latter has limitations such as the potential development of an immune response and the risk of disease transfer. Tissue engineering has emerged as a multidisciplinary field that comprises principles from biology, chemistry, medicine and engineering [38, 39] towards the goal of tissue regeneration, tissue restoration, maintenance or improvement of tissue function.

The main tissue engineering strategy is depicted in Figure 4: a biological component, primarily made of harvested cells is expanded in vitro for days or weeks. Then cells are seeded onto scaffolds and cultured for different time intervals, ranging form days to months, depending on the tissue to be regenerated. The 3D hybrid tissue thus obtained is subsequently implanted at the patient's lesion site. Tissue engineering relies mainly on the development of scaffolds able to mimic the tissue architecture at the nanoscale, which are seeded with cells derived ideally from the patient biopsy.

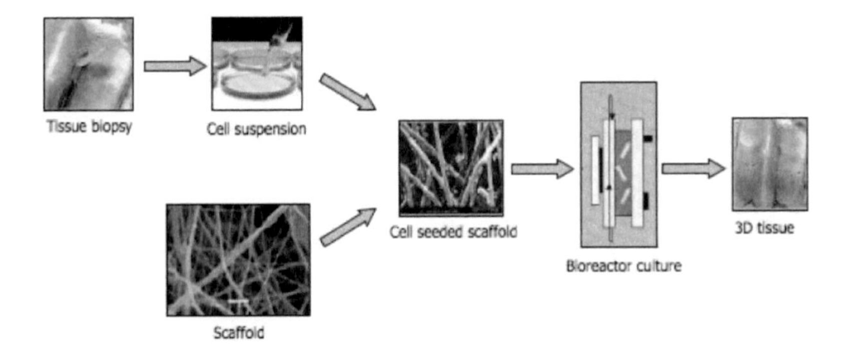

Figure 4. General tissue engineering strategy. Modified from Rose and Oreffo [57].

These scaffolds need to satisfy a number of conditions, such as: biocompatibility and absence of immune response and bioresorbability and/or biodegradability. In addition, they should present characteristics in accordance to the characteristics of the tissue to be replaced, such as suitable mechanical properties, suitable porosity and morphology and suitable physicochemical properties. Using electrospinning, researchers have been tailoring different scaffolds to meet the requirements of the tissue to be engineered. We will here give an overview of the most recent electrospun biomaterials tested either *in vitro* or *in vivo* for tissue engineering applications.

3.2. Musculoskeletal Tissue Engineering

Though nanofibers have been studied as scaffolds for several tissue types, musculoskeletal tissue is almost certainly the most well studied one.

Skeletal tissue engineering requires, essentially, a scaffold conducive to cell attachment and maintenance of cell function, together with a rich source of osteoprogenitor cells in combination with selected osteoinductive growth factors.

3.2.1. Bone

The requirement for new bone to replace or restore the function of traumatised, damaged or lost bone is a major clinical and socioeconomic need.

The natural bone is composed of a hierarchical distribution of collagen and vital minerals (mainly calcium phosphate). Especially at the bone-cartilage interface both the concentrations and the orientations of the collagen and calcium phosphate particles are precisely organized. The engineering of a fully functional bone tissue still remains elusive, despite the excellent progress achieved up to date.

The typical composite scaffold consists of a biodegradable polymer homogeneously incorporated with various additives including tricalcium phosphate, hydroxyapatite, calcium carbonate, carbon nanotubes, hydrogels and proteins.

Bone formation was assessed in a rat model, by seeding mesenchymal stem cells on a PCL electrospun scaffold presenting an ECM-like topography. The cell-polymer constructs were cultured with osteogenic supplements in a rotating bioareactor for 4 weeks and subsequently implanted in the omenta of rats for 4 weeks. After explantation, the constructs presented a rigid and bone-like appearance and mineralization and type I collagen were detected [39]. An electrospun silk fibroin membrane seeded with a mouse preosteoblast cell line was analysed *in vivo* in a rabbit model with a 8 mm bilateral full-thickness calvarial bone defect. The prosthesis was shown to possess good biocompatibility and to effectively enhance new bone formation in vivo within 12 weeks [40].

A hybrid twin-screw extrusion/electrospinning process, which generates continuous spatial gradations in composition and porosity of nanofibers, was recently developed for the fabrication of non woven meshes of PCL incorporated with ß-tricalcium phosphate, to be used in the area of bone tissue regeneration and especially towards the controlled formation of the bone-cartilage interface.

The scaffolds were seeded with mouse preosteoblast cells and within 4 weeks the tissue constructs revealed the formation of continuous gradations in ECM with various markers including collagen synthesis and mineralization, with resemblance to the type of variations observed in the typical bone-cartilage interface in terms of the distributions of concentration of Ca^{+} particles and of mechanical properties associated with this [12].

3.2.2. Cartilage

Unlike bone, which has shown some prowess for repair and even regeneration, cartilage is recalcitrant to repair, mostly due to its hierarchical organization and geometry. In fact, cartilage presents a very complex stratified tissue structure. Cartilage is composed mainly of type II collagen, so that electrospinning of type II collagen was performed and scaffolds demonstrated to produce a suitable environment for chondrocyte growth which potentially establishes the foundation for the development of articular cartilage repair [15].

3.3. Skin Tissue Engineering

The complex nature of wound healing requires the migration and proliferation of keratinocytes, both phenomena temporally-regulated by numerous growth factors activating cell membrane receptors up-regulated in the wound environment [41, 42].

Tissue engineering projects innovative scaffolds to promote the adhesion and proliferation of human skin fibroblasts and keratinocytes,

Electrospun poly(lactic acid-co-glycolic acid (PLGA) matrices with fiber diameters from 150 to 6000 nm were fabricated and tested for their efficacy as skin substitutes by seeding them with human skin fibroblasts (hSF). hSF acquired a well spread morphology and showed significant progressive growth on fiber matrices in the 350-1100 nm diameter range [43].

An electrospun fibrinogen scaffold was cross-linked with one of the three cross-linkers: glutaraldehyde vapour, 1-ethyl-3-(3-dimethylaminopropyl) carbodiimide hydrochloride (EDC) in ethanol and genipin in ethanol. All three cross-linked scaffolds were seeded with human foreskin fibroblasts. EDC and genipin in ethanol proved to be highly effective in enhancing scaffold mechanical properties and in retarding the rate of scaffold degradation, in respect to the non cross-linked scaffold. Yet, this study demonstrated also that these cross-linked scaffolds had a negative impact on the ability of fibroblasts to migrate below the surface of the scaffold and remodel it with collagen [44].

3.4. Vascular Tissue Engineering

The vascular tissue arrangement precedes and dictates the architecture of the new tissue to be engineered, so that we have to consider both the question of vascular tissue engineering *per se* and also as a condition for musculoskeletal, skin and neural tissue engineering.

Making a selection of materials to be electrospun for arterial blood vessels, the energy and shape recovery are critical parameters to be considered. Energy stored during the expansion of the blood vessel should be recoverable and used in the contraction of the vessel without any distortion to the vessel.

An attractive option of using electrospinning to fabricate vascular grafts is its ability to electrospin small diameter tubes of different sizes with uniform thickness and fiber distribution throughout the scaffold. In fact, many vascular graft scaffolds have been fabricated. An electrospun polymer blend of type I collagen, elastin, and PLGA was used to fabricate a tubular scaffold of 4.75 mm inner diameter.

The scaffolds were shown to be biocompatible and to possess tissue composition and mechanical properties similar to native vessels. Also, it was found to support both smooth muscle cells and endothelial cells [45]. The constructed vascular grafts should moreover express anti-coagulant activity until the endothelial cell lining is fully achieved.

A solution blend of PLCL and a tri-n-butylamine salt of heparin was electrospun. Its soaking in PBS determined a burst release of heparin in the first 12 hours, after which relatively sustained release rate was observed for 4 weeks [46].

3.5. Cardiac Tissue Engineering

In the context of heart valve engineering, it is emphasized the need for including the requirements derived from "adult biology" of tissue remodelling and establishing reliable early predictors of success or failure of tissue engineered implants. In the fabrication of cardiac graft, electrospun PCL scaffolds were coated with purified type I collagen solution to promote cell attachment. Neonatal rat cardiomyocytes were cultured on the electrospun PCL scaffolds [47].

The cardiomyocytes attached well to the scaffold and contraction of the cardiomyocytes was observed. Tight arrangement and intercellular contacts of the cardiomyocytes were formed throughout the entire mesh, although more cells were found on the surface. The electrospun scaffold was sufficiently soft such that contractions of the cardiomyocytes were not impeded and stable enough for handling. By suspending the mesh across a ring, the cardiomyocytes are allowed to contract at their natural frequency. Thus electrospun patches seeded with cardiomyocytes are gaining interest among the scientific community for the recovery of infarctuated myocradia.

The use of electrical stimulations has been shown to increase adsorption of serum proteins onto electrically conducting polymer, which leads to significantly enhanced neurite extension [48].

In developing nano-fibrous scaffolds to modulate various cell functions such as proliferation, differentiation and migration through electrical simulation, a blend of polyaniline and gelatin was electrospun and it was found to be biocompatible, supporting attachment, migration and proliferation of H9c2 rat cardiac myoblasts [49].

3.6. Neural Tissue Engineering

Engineering the neural tissue would be important for a number of applications, ranging from neural probes for neurodegenerative diseases to guidance scaffolds for axonal regeneration in patients with traumatic nerve injuries. Neural injury may be treated more effectively using nerve guidance channels containing longitudinally aligned fibers and this is true for both the PNS and the spinal cord.

It is well demonstrated that aligned electrospun PLLA nanofibers direct NSC neurite outgrowth, being a good candidate to be used as a potential scaffold in neural tissue engineering [50]. Also, PLLA electrospun nanofiber were shown to support the serum-free growth of primary motor and sensory neurons [51]. Also PLC/chitosan electrospun nanofibers demonstrated good results on Schwann cell proliferation and maintenance of cell morphology, with spreading bipolar elongations to the nanofibrous substrates [52]. Also, a copolymer of methyl methacrylate and acrylic acid was electrospinned and cultured with neural stem cells and it was demonstrated that when type I collagen was immobilized onto the nanofibers surface, cell attachment and viability was enhanced [53].

In vivo studies have been performed and recently an electrospun guidance channel made of a blend of PLGA and PCL was used to regenerate a 10-mm nerve gap in a rat model of sciatic nerve transaction, with no additional biological coating or drug loading treatment [54]. Also, an electrospun bilayered chitosan tube was fabricated, comprising an outer layer of chitosan film and an inner layer of chitosan nonwoven nano and microfiber mesh and moreover, the inner layer of the tube was covalently bound with peptides with modified domains for laminin-1. This tube was grafted to bridge a rat injured sciatic nerve and nervous regeneration obtained was similar to the control isograft [55]. Moreover, it was demonstrated that chitosan nano and microfiber mesh tubes with a deacetylation rate of 93% present good nerve regeneration in a rat sciatic nerve injury model [56].

CONCLUSION

Electrospinning is with no doubt a technique which much potential for different applications. Due to its versatility, it can be applied in different fields as scaffolds for tissue engineering, protective clothing, reinforcement in composite materials and sensors. This chapter focus only in the specific application of the electrospinning technique in the field of biomedicine, but as much chapters could be written for each of the referred areas.

We have particularly focused on tissue engineering applications, an area which by its nature produces an enormous amount of new data each year and much are due to the use of electrospinning for the design of biomaterials for cell growth. Tissue engineering aims at mimicking the particular ECM of the tissue to be engineering and in doing so its counts mainly on the development of nanofibers. These allow the development of scaffolds with high surface area/volume ratio and enhanced porosity, properties that demonstrated to have a significant effect on cell adhesion, proliferation and differentiation. Nanofibers can be now fabricated from a large range of natural and synthetic biomaterials by different techniques but has been electrospinning the one that has shown the most promising results and the most widely studied. It is a tightly regulated process, which presents high versatility not only in the materials that can be fabricated but also in the numerous modification possibilities on their characteristics. In this chapter, the most recent advances in this area have been reviewed and as we saw, electrospinning has effectively contributed for the advancement of muscoloskeletal (including bone, cartilage, ligament, and skeletal muscle), skin, vascular, cardiac and neural tissue engineering applications. In the future, electrospun engineered tissue will have to be not only biomimetic but also bioactive, so that fibers will have to be modified to incorporate and to release at specific rates drugs, proteins and DNA, which as we saw are fundamental for the correct design of tissue engineering. Electrospinning is for now the most advanced technique of this century for tissue engineering applications and will probably lead the way for effectively producing a clinically useful engineered tissue substitute.

List of Abbreviations

3D three-dimensional
DNA deoxyribonucleic acid
ECM extracellular matrix
hSF human skin fibroblasts
NSCs neural stem cells
PBS phosphate buffered saline
PCL policaprolactone
PGA poly(glycolic acid)

PLGA poly(lactic-co-glycolic acid)
PLLA poly(L-lactic acid)
TFA trifluoroacetic acid

ACKNOWLEDGMENTS

FG, SP and CC gratefully acknowledge Cariplo Foundation and Regione Lombardia for their grant support.

REFERENCES

[1] Jayaraman K, Kotaki M, Zhang Y, Mo X, Ramakrishna S. Recent advances in polymer nanofibers. *J. Nanosci. Nanotechnol.* 2004 Jan-Feb;4(1-2):52-65.
[2] Cooley JF, inventor. *US patent* 692,631. 1902.
[3] Morton WJ, inventor. *US patent* 705,691. 1902.
[4] Formhals J, inventor. *US patent* 2,077,373. 1934.
[5] Formhals J, inventor. *US patent* 2,158,415. 1939.
[6] Formhals J, inventor. *US patent* 2,323,02. 1943.
[7] Formhals J, inventor. *US patent* 2,349,950. 1944.
[8] Norton CL, inventor. *US patent* 2,048,651. 1936.
[9] Theron A, Zussman E, Yarin AL. Electrostatic field-assisted alignment of electrospun nanofibres. *Nanotechnology*. 2001;12:384-90.
[10] Kidoaki S, Kwon IK, Matsuda T. Mesoscopic spatial designs of nano- and microfiber meshes for tissue-engineering matrix and scaffold based on newly devised multilayering and mixing electrospinning techniques. *Biomaterials*. 2005 Jan;26(1):37-46.
[11] Sun T, Mai S, Norton D, Haycock JW, Ryan AJ, MacNeil S. Self-organization of skin cells in three-dimensional electrospun polystyrene scaffolds. *Tissue engineering*. 2005 Jul-Aug;11(7-8):1023-33.
[12] Erisken C, Kalyon DM, Wang H. Functionally graded electrospun polycaprolactone and beta-tricalcium phosphate nanocomposites for tissue engineering applications. *Biomaterials*. 2008 Oct;29(30):4065-73.
[13] Lannutti J, Reneker D, Ma T, Tomasko D, Farson D. Electrospinning for tissue engineering scaffolds. *Materials Science and Engineering*. 2007;C(27):504-09.

[14] Yeo IS, Oh JE, Jeong L, Lee TS, Lee SJ, Park WH, et al. Collagen-based biomimetic nanofibrous scaffolds: preparation and characterization of collagen/silk fibroin bicomponent nanofibrous structures. *Biomacromolecules*. 2008 Apr;9(4):1106-16.

[15] Shields KJ, Beckman MJ, Bowlin GL, Wayne JS. Mechanical properties and cellular proliferation of electrospun collagen type II. *Tissue engineering*. 2004 Sep-Oct;10(9-10):1510-7.

[16] Khanam N, Mikoryak C, Draper RK, Balkus KJ, Jr. Electrospun linear polyethyleneimine scaffolds for cell growth. *Acta biomaterialia*. 2007 Nov;3(6):1050-9.

[17] Li Y, Ma T, Yang ST, Kniss DA. Thermal compression and characterization of three-dimensional nonwoven PET matrices as tissue engineering scaffolds. Biomaterials. 2001 MAR;22(6):609-18.

[18] Li WJ, Tuli R, Huang X, Laquerriere P, Tuan RS. Multilineage differentiation of human mesenchymal stem cells in a three-dimensional nanofibrous scaffold. *Biomaterials*. 2005 Sep;26(25):5158-66.

[19] Bhattarai SR, Bhattarai N, Viswanathamurthi P, Yi HK, Hwang PH, Kim HY. Hydrophilic nanofibrous structure of polylactide; fabrication and cell affinity. *Journal of biomedical materials research*. 2006 Aug;78(2):247-57.

[20] Telemeco TA, Ayres C, Bowlin GL, Wnek GE, Boland ED, Cohen N, et al. Regulation of cellular infiltration into tissue engineering scaffolds composed of submicron diameter fibrils produced by electrospinning. *Acta biomaterialia*. 2005 JUL;1(4):377-85.

[21] Li WJ, Laurencin CT, Caterson EJ, Tuan RS, Ko FK. Electrospun nanofibrous structure: a novel scaffold for tissue engineering. *J. Biomed. Mater Res*. 2002 Jun 15;60(4):613-21.

[22] Wang Y, Ameer GA, Sheppard BJ, Langer R. A tough biodegradable elastomer. *Nature biotechnology*. 2002 Jun;20(6):602-6.

[23] Larrondo L, Manley RSJ. Electrostatic Fiber Spinning From Polymer Melts. 1.Experimental Observations on Fiber Formation and Properties. *Journal of Polymer Science: Polymer Physics Edition*. 1981;19:909-20.

[24] Larrondo L, Manley RSJ. Electrostatic Fiber Spinning From Polymer Metls. 2.Examination of the Flow Field in an Electrically Driven Jet. *Journal of Polymer Science: Polymer Physics Edition*. 1981;19:921-32.

[25] Li C, Vepari C, Jin HJ, Kim HJ, Kaplan DL. Electrospun silk-BMP-2 scaffolds for bone tissue engineering. *Biomaterials*. 2006 Jun;27(16):3115-24.

[26] Kim TG, Park TG. Biomimicking extracellular matrix: cell adhesive RGD peptide modified electrospun poly(D,L-lactic-co-glycolic acid) nanofiber mesh. *Tissue engineering*. 2006 Feb;12(2):221-33.
[27] Fridrikh SV, Yu JH, Brenner MP, Rutledge GC. Controlling the fiber diameter during electrospinning. *Physical review letters*. 2003 Apr 11;90(14):144502.
[28] Simonet M, Schneider OD, Neuenschwander P, Stark WJ. Ultraporous 3D polymer meshes by low-temperature electrospinning: Use of ice crystals as a removable void template. *Polymer Engineering and Science*. 2007 DEC;47(12):2020-6.
[29] Nam J, Huang Y, Agarwal S, Lannutti J. Improved cellular infiltration in electrospun fiber via engineered porosity. *Tissue engineering*. 2007 Sep;13(9):2249-57.
[30] Jena A, Gupta K. Liquid Extrusion Techniques For Pore Structure Evaluation Of Nonwovens. *International Nonwovens Journal*. 2003:45-55.
[31] Jena A, Gupta K. Pore Volume of Nanofiber Nonwovens. *International Nonwovens Journal*. 2005:25-30.
[32] Lee YH, Lee JH, An IG, Kim C, Lee DS, Lee YK, et al. Electrospun dual-porosity structure and biodegradation morphology of Montmorillonite reinforced PLLA nanocomposite scaffolds. *Biomaterials*. 2005 Jun;26(16):3165-72.
[33] Stupp SI, LeBonheur VV, Walker K, Li LS, Huggins KE, Keser M, et al. Supramolecular Materials: Self-Organized Nanostructures. *Science*. 1997 Apr 18;276(5311):384-9.
[34] Zhang R, Ma PX. Synthetic nano-fibrillar extracellular matrices with predesigned macroporous architectures. *J. Biomed. Mater Res*. 2000 Nov;52(2):430-8.
[35] Hogge J, Krasner D, Nguyen H, Harkless LB, Armstrong DG. The potential benefits of advanced therapeutic modalities in the treatment of diabetic foot wounds. *J. Am. Podiatr. Med. Assoc*. 2000 Feb;90(2):57-65.
[36] O'Meara S, Cullum N, Majid M, Sheldon T. Systematic reviews of wound care management: (3) antimicrobial agents for chronic wounds; (4) diabetic foot ulceration. *Health Technol. Assess*. 2000;4(21):1-237.
[37] Menzies D. Peritoneal adhesions. Incidence, cause, and prevention. *Surg. Annu.* 1992;24 Pt 1:27-45.
[38] Langer R, Vacanti JP. Tissue engineering. *Science*. 1993 May 14;260(5110):920-6.

[39] Shin M, Yoshimoto H, Vacanti JP. In vivo bone tissue engineering using mesenchymal stem cells on a novel electrospun nanofibrous scaffold. *Tissue engineering*. 2004 Jan-Feb;10(1-2):33-41.

[40] Kim KH, Jeong L, Park HN, Shin SY, Park WH, Lee SC, et al. Biological efficacy of silk fibroin nanofiber membranes for guided bone regeneration. *J. Biotechnol*. 2005 Nov 21;120(3):327-39.

[41] Ono I, Gunji H, Zhang JZ, Maruyama K, Kaneko F. Studies on cytokines related to wound healing in donor site wound fluid. *Journal of dermatological science*. 1995 Nov;10(3):241-5.

[42] Tamariz-Dominguez E, Castro-Munozledo F, Kuri-Harcuch W. Growth factors and extracellular matrix proteins during wound healing promoted with frozen cultured sheets of human epidermal keratinocytes. *Cell and tissue research*. 2002 Jan;307(1):79-89.

[43] Kumbar SG, Nukavarapu SP, James R, Nair LS, Laurencin CT. Electrospun poly(lactic acid-co-glycolic acid) scaffolds for skin tissue engineering. *Biomaterials*. 2008 Oct;29(30):4100-7.

[44] Sell SA, Francis MP, Garg K, McClure MJ, Simpson DG, Bowlin GL. Cross-linking methods of electrospun fibrinogen scaffolds for tissue engineering applications. *Biomed. Mater*. 2008 Sep 25;3(4):45001.

[45] Stitzel J, Liu J, Lee SJ, Komura M, Berry J, Soker S, et al. Controlled fabrication of a biological vascular substitute. *Biomaterials*. 2006 Mar;27(7):1088-94.

[46] Kwon IK, Matsuda T. Co-electrospun nanofiber fabrics of poly(L-lactide-co-epsilon-caprolactone) with type I collagen or heparin. *Biomacromolecules*. 2005 Jul-Aug;6(4):2096-105.

[47] Shin M, Ishii O, Sueda T, Vacanti JP. Contractile cardiac grafts using a novel nanofibrous mesh. *Biomaterials*. 2004 Aug;25(17):3717-23.

[48] Kotwal A, Schmidt CE. Electrical stimulation alters protein adsorption and nerve cell interactions with electrically conducting biomaterials. *Biomaterials*. 2001 May;22(10):1055-64.

[49] Li M, Guo Y, Wei Y, MacDiarmid AG, Lelkes PI. Electrospinning polyaniline-contained gelatin nanofibers for tissue engineering applications. *Biomaterials*. 2006 May;27(13):2705-15.

[50] Yang F, Murugan R, Wang S, Ramakrishna S. Electrospinning of nano/micro scale poly(L-lactic acid) aligned fibers and their potential in neural tissue engineering. *Biomaterials*. 2005 May;26(15):2603-10.

[51] Corey JM, Gertz CC, Wang BS, Birrell LK, Johnson SL, Martin DC, et al. The design of electrospun PLLA nanofiber scaffolds compatible with

serum-free growth of primary motor and sensory neurons. *Acta biomaterialia*. 2008 Jul;4(4):863-75.

[52] Prabhakaran MP, Venugopal J, Chyan TT, Hai LB, Chan CK, Tang AL, et al. Electrospun Biocomposite Nanofibrous Scaffolds for Neural Tissue Engineering. *Tissue Eng Part A*. 2008 Jul 25.

[53] Li W, Guo Y, Wang H, Shi D, Liang C, Ye Z, et al. Electrospun nanofibers immobilized with collagen for neural stem cells culture. *J. Mater Sci. Mater Med*. 2008 Feb;19(2):847-54.

[54] Panseri S, Cunha C, Lowery J, Del Carro U, Taraballi F, Amadio S, et al. Electrospun micro- and nanofiber tubes for functional nervous regeneration in sciatic nerve transections. *BMC Biotechnol*. 2008;8:39.

[55] Wang W, Itoh S, Matsuda A, Aizawa T, Demura M, Ichinose S, et al. Enhanced nerve regeneration through a bilayered chitosan tube: the effect of introduction of glycine spacer into the CYIGSR sequence. *Journal of biomedical materials research*. 2008 Jun 15;85(4):919-28.

[56] Wang W, Itoh S, Matsuda A, Ichinose S, Shinomiya K, Hata Y, et al. Influences of mechanical properties and permeability on chitosan nano/microfiber mesh tubes as a scaffold for nerve regeneration. *Journal of biomedical materials research*. 2008 Feb;84(2):557-66.

[57] Rose FR, Oreffo RO. Bone tissue engineering: hope vs hype. *Biochemical and biophysical research communicati*ons. 2002 Mar 22;292(1):1-7.

In: Electrospinning Process and Nanofiber... ISBN 978-1-61209-330-7
Editors: A.K. Haghi and G.E. Zaikov

Chapter 9

WORKING WITH ELECTROSPUN SCAFFOLDS: SOME PRACTICAL HINTS FOR TISSUE ENGINEERS

Maria Letizia Focarete [1], Chiara Gualandi [1] and Lorenzo Moroni [2]

[1]Department of Chemistry G Ciamician, University of Bologna, Via Selmi 2, 40126 Bologna - Italy

[2] Muscoloskeletal Tissue Bank, Rizzoli Orthopaedic Institute, Via G.C. Pupilli, 1, 40136 Bologna – Italy

ABSTRACT

Polymer non-woven mats are often considered as potential three-dimensional (3D) supports (scaffolds) for tissue engineering applications, where cells and bioactive molecules are combined with a proper scaffold to repair and regenerate damaged biological tissues. Electrospinning is a promising technology for the fabrication of nanofibrous non-woven mats that resemble the morphological nano-features of the extracellular matrix (ECM). For this reason, electrospun meshes are widely used as ECM-mimicking scaffolds to enhance cell-material interactions and tissue regeneration. In order to properly use electrospun scaffolds it is important to take into account on one hand the well-known problem of electrospinning process reproducibility and, on the other hand, all practical aspects related with scaffold handling and scaffold preparation for cell culture experiments. As a matter of fact, in some cases the above mentioned issues can dramatically change fibre morphology, that is known to affect viability, attachment and migration of cells seeded on the scaffold.

In this chapter the reproducibility of the electrospinning process will be discussed and practical hints, concerning for example wetting procedure, scaffold sterilization, mat shrinkage, scaffold handling, etc., will be provided to tissue engineers using electrospun scaffolds in cell culturing experiments. It will be also pointed out that a proper understanding of polymeric solid state properties is required in order to improve standard operating procedures to manufacture electrospun scaffolds for regenerative medicine use.

1. INTRODUCTION

Polymeric scaffolds made through electrospinning technology have found widespread applications in tissue engineering, a multidisciplinary field that aims at repairing damaged tissues and restoring their functionalities in the body. To achieve this, cells obtained by a patient biopsy are cultured through *in vitro* techniques and seeded on a natural or synthetic support that acts as a temporary three-dimensional (3D) scaffold, replacing the natural extra-cellular matrix (ECM) [1-5]. 3D scaffolds can be fabricated by a large number of technologies [5-11]. Among these, electrospinning has acquired increasing interest as it allows generating porous non-woven scaffolds comprised of micro and nanofibers, that resemble ECM physical structure [12-14]. Fibre morphology and fibre deposition pattern can be tailored to specific design criteria by controlling both the experimental parameters and the instrumental apparatus configuration (e.g. type of collector, counter fields, etc.) [15-18]. Electrospun meshes can be fabricated from solutions of a single polymer or of different polymers by co-axial or layer-by-layer spinning strategies [19,20]. The resulting electrospun ECM-inspired scaffolds have demonstrated to support enhanced cell adhesion, proliferation and differentiation when compared to 3D scaffolds with macro-scaled dimensions [21-23]. Further functionalization of the scaffold can be easily obtained by combining polymer solutions with growth factors and cell suspensions, thereby increasing cell viability and the quality of the formed neo-tissue [24-27]. Exploiting all these possibilities makes electrospinning a very promising scaffold fabrication technique for successful applications in the regeneration of vessels, skin, and other soft connective tissues [28-31], as well as of stiffer tissues like bone and cartilage [27,32]. In addition, electrospinning can be easily combined with other fabrication technologies to create multifunctional structures that display adequate mechanical and physico-chemical properties [33,34]. Even if electrospinning is nowadays an established scaffold fabrication platform, many issues have to be taken into account to properly use electrospun scaffolds in tissue engineering applications. As regards scaffold fabrication, it will be demonstrated in the following paragraphs that an accurate control of environmental parameters is a crucial condition in order to obtain product reproducibility. Other pitfalls that are commonly encountered during manipulation of electrsopun scaffolds in tissue engineering can be circumvented by practical tricks. The choice of the most appropriate scaffold sterilization technique, for example, is a critical issue strictly depending on the physical properties of the polymeric material. Removal of the air entrapped in the pores before cell seeding on electrsopun mats is also a key step for successful applications. If this is not completely achieved cells as well as all substances and metabolites that have to diffuse through the scaffold will find physical barrier for migration. Another problem that will be discussed in the following paragraphs is the dimensional stability of the electrospun scaffolds that, in some cases, is not maintained during cell culture experiments. Although these issues are often the subject of practical optimization in the lab, possible solutions are rarely discussed in the literature. In the present chapter we will illustrate some practical hints, deriving from a deep knowledge of polymeric chemical-physical properties, that may be useful to improve handling of electrospun scaffolds.

2. Polymer Physical-Chemical Properties

Synthetic bioresorbable polymers are largely employed for tissue engineering applications. They posses a wide range of chemical, physical, mechanical and degradation properties, depending primarily on their chemical structure and molecular weight. Moreover, properties can be tailored through copolymerization of two different monomers or blending of two different polymers. Therefore, such polymeric systems are very versatile materials for the design of scaffolds with properties approaching the specific structural and functional requirements of the natural tissues to be regenerated. A careful selection of a proper material must be accompanied by the correct choice of the technology used to fabricate a scaffold with the desired 3D structure. As regards electrospinning technology, the chemical and physical properties of a polymer can, in some cases, limit its processability into fibrous meshes. For instance, it is well known that high molecular weight polymers are more easily electrospun than low molecular weight ones. Indeed, several studies demonstrate that fiber formation and fiber morphology are strictly correlated with the number of chain entanglements in the polymeric solution [35,36], which depends primarily on polymer molecular weight. Electrospinning process feasibility is also related to polymer solid state properties that depend on several factors, the most significant ones being: (i) the chemical structure of the repeating unit, (ii) the molecular weight and molecular weight distribution, (iii) the presence of chain branching, and (iv) the nature of inter- and intramolecular chain interactions. Depending on the regularity of their repeating units, macromolecules can arrange themselves in an ordered structure, thus developing crystal regions, or they can form a disordered amorphous phase. In the amorphous regions, macromolecules are in a glassy or in rubbery state depending on their mobility. The transition from the glassy to the rubbery state and vice versa is called glass transition. When thermally characterized, completely amorphous polymers only exhibit the glass transition (at a temperature indicated as T_g), whereas semicrystalline polymers - containing both amorphous and crystalline regions - exhibit also the fusion of the crystal phase (at a temperature T_m). Thermal properties of the most common synthetic bioresorbable polymers that have been successfully electrospun for tissue engineering applications are listed in Table 1 together with the physical state of the obtained electrospun scaffolds.

Table 1 shows that the polymers that have been successfully electrospun either have a T_g higher than room temperature (RT), when completely amorphous (e.g. P(D,L)LA), or they possess a high amount of crystal phase, when their T_g is lower than RT (e.g. PCL). This finding can be explained considering that fibres obtained by electrospinning maintain a stable morphology at RT only when the polymer chain mobility is very low ($T_g >$ RT, i.e. macromolecules in glassy state) or when a crystalline phase, acting as structural supporting phase, is present. As an example, in our laboratory we electrospun a copolymer of poly(3-hydroxybutyrate-co-3-hydroxyhexanoate), P(3HB-*co*-3HH), (3HH content = 13 mol%), characterized by a T_g around 2 °C and by a multiple melting endotherm in the range 50 - 130 °C. During the electrospinning process, this polymer developed a very low amount of crystal phase that was not enough to obtain a stable fibre morphology. Indeed, fibres collapsed as soon as reaching the collector, because of the very high polymer chain mobility of the rubbery amorphous phase and a compact "film-like" structure was obtained instead of the expected fibrous mat.

Table 1. Thermal properties of bioresorbable polymers commonly used as electrospun materials in tissue engineering[a] and physical state of electrospun scaffolds.

Polymer	Tg (°C)	Tm (°C)	Physical state of electrospun scaffold
polyglycolic acid [PGA]	35	210	Semicrystalline [37-39]
poly(L-lactic acid) [P(L)LA]	55 ÷ 60	159 ÷ 185	Amorphous [40,41] Semicrystalline [42,43]
poly(D,L-lactic acid) [P(D,L)LA]	45 ÷ 55	-	Amorphous [41]
poly(D,L-lactide-co-glycolide) [P(LA-co-GA)]	37 ÷ 55 [b]	180 ÷ 200 [b]	Amorphous [38, 44] Semicrystalline [45]
poly(ε-caprolactone) [PCL]	-62	57	Semicrystalline [46]
Poly-3-hydroxybutyrate [PHB]	1	171	Semicrystalline [47,48]
Poly(3-hydroxybutyrate -co-3-hydroxyvalerate) [P(HB-co-HV)]	-5 ÷ -1 b)	137 ÷ 160 b)	Semicrystalline [47]

[a]from references [49-52].

[b]depending on copolymer molar composition.

Understanding material properties is a necessary requirement, not only in order to predict polymer processability, but also to better operate with fabricated electrospun scaffolds. As an example, familiarity with polymer properties helps in modifying the polymer crystallinity degree. It is well known that the amount of crystal phase strongly influences mechanical properties and degradation mechanism of a scaffold. P(L)LA is a polymer able to crystallize thanks to the structural regularity of its chains, but its electrospun scaffolds are usually either completely amorphous [40,41] or they contain a very low amount of crystallinity [42,43]. Indeed, it has been hypothesized that the electrospinning process inhibits crystallization [41,45,53].

Figure 1.a. reports a scanning electron microscopy (SEM) micrograph of a completely amorphous as-spun P(L)LA scaffold (thermal analysis not shown). However, if desired, it is possible to induce the formation of a crystal phase in P(L)LA electrospun fibres by a proper post-treatment. The common strategy applied to this aim is annealing the polymer at a constant crystallization temperature, located between T_g and T_m (crystallization window). Figure 1.b. depicts SEM micrographs of a P(L)LA electrospun scaffold subjected to an annealing treatment at 70 °C for 1 h. Thermal characterization (data not shown) demonstrates that the annealed mat is semicrystalline. Careful inspection of the micrographs in Figure 1.a. and Figure 1.b. shows however that a high temperature treatment is not recommended for electrospun scaffolds, because fibres can undergo dimensional deformation, flattening, and collapsing. An alternative strategy in order to induce crystallization in the amorphous as-spun scaffold is the treatment of the mat with a solvent that acts as a plasticizer agent for the polymer. The plasticizer decreases polymer T_g and thus extends the crystallization window, allowing crystallization to occur at a lower temperature. Indeed, thermal analysis of P(L)LA sample soaked in ethanol (EtOH) reveals that the polymer T_g decreases from 63 °C to 22 °C. This plasticization effect allows the P(L)LA elctrospun mat to crystallize when kept in EtOH at 35 °C overnight. Figure 1.c. shows SEM micrograph of a P(L)LA fibre mat after EtOH treatment. Comparing Figure 1.a. with Figure 1.c. unveils that the treatment with EtOH does not significantly affect fibre morphology, in contrast with the result obtained with the annealing treatment.

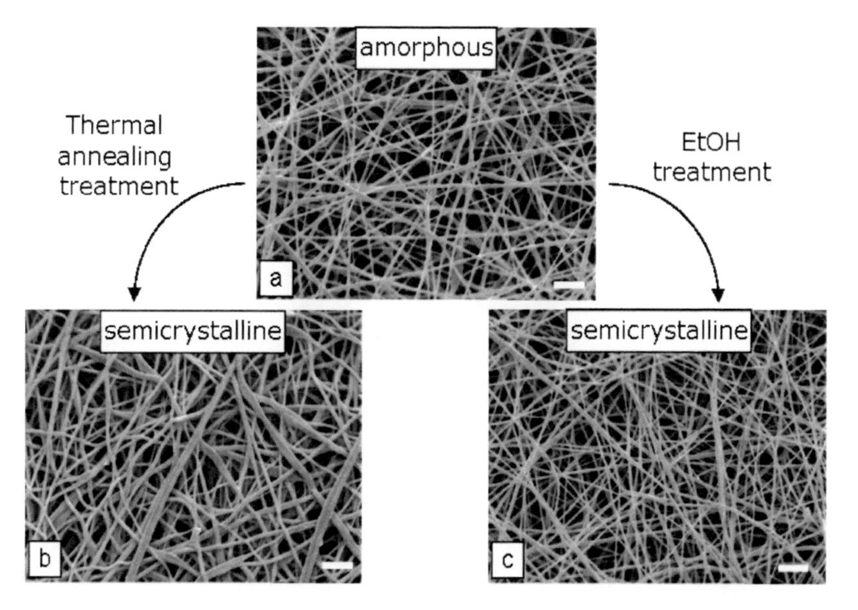

Figure 1.a) P(L)LA as-spun fibres, b) P(L)LA fibres after thermal annealing and c) P(L)LA fibres after EtOH treatment. Scale bar = 10 μm.

3. CONTROL OF ENVIRONMENTAL CONDITIONS

Electrospun fibre morphology - i.e. fibre diameter, presence of bead defects and fibre surface topography - depends on process variables, which can be divided into three categories: (i) solution parameters (molecular weight and concentration of the polymer; electrical properties, boiling point and surface tension of the solution); (ii) instrumental parameters (applied voltage, needle to collector distance, solution flow rate, needle diameter) and (iii) environmental parameters (temperature and relative humidity) [54,55]. Although solution parameters and instrumental variables can be easily set, control of environmental conditions is rather difficult to obtain, since it requires the equipment to be placed in a conditioned box. This kind of facility is rarely included in conventional laboratory-scale electrospinning setup. As a consequence, ambient parameters are often neglected, despite their strong influence on the resulting fibre morphology. For this reason, electrospinning the same polymer solution by using the same process parameters but in different environmental conditions often lead to non-reproducible fibre morphology. For solution electrospinning, the influence of temperature on fibre morphology – not often investigated in the literature - has been related to its effect on solvent evaporation rate and on solution viscosity [56].

Conversely, many studies are reported on the effect of humidity on fibre morphology. In particular, it is well documented that humidity influences surface fibre porosity, due to phase separation phenomena or to the condensation of water drops on the fibre during solvent evaporation [57-59]. The effect of humidity on fibre diameter and on the occurrence of beads is however a less studied topic [56,60-62]. Indeed, the influence of humidity on fibre morphology is a rather complex issue, since this parameter can have several concomitant effects on electrospinning process. First, water molecules in the region between the needle and the collector are oriented by the electric field, with the effect of modifying, to some extent, the neat electric field that controls jet elongation. Second, in the presence of high humidity, during the travel towards the collector the polymer jet easily looses its charges, thus limiting its elongation and stretching.

Moreover, water molecules can be absorbed by the polymer jet, thus inducing a faster solidification of the jet if water is a non-solvent for the electrospun polymer. Humidity can also influence the rate of solvent evaporation, especially for polymers electrospun from water solutions. All these effects come into play to control fibre morphology and the overall result can be different when different systems and different processing conditions are used. Therefore, rationalization of humidity effect in the electrospinning process is rather difficult to achieve.

However, it is important to take into account that small changes of environmental humidity can sometimes remarkably affect reproducibility of fibre morphology, as it is illustrated by the following examples.

Example 1. A 11% (wt/v) solution of P(L)LA (M_w = 8.4 x 10^4 g/mol, PDI = 1.7) in 65:35 v/v dichloromethane:dimethylformamide was electrospun on an aluminium plate. The process was performed in a glove box under controlled temperature and humidity conditions: the temperature was maintained at (20 ± 1) °C and the relative humidity (RH) was changed in order to evaluate its effect on fibre morphology. SEM images of P(L)LA fibres obtained at different humidity conditions are reported in Figure 2. At low humidity many beads are present along the fibres (Figure 2.a), while increasing relative humidity from 30% to 60% results in a remarkable decrease of bead number (Figures 2.b-d).

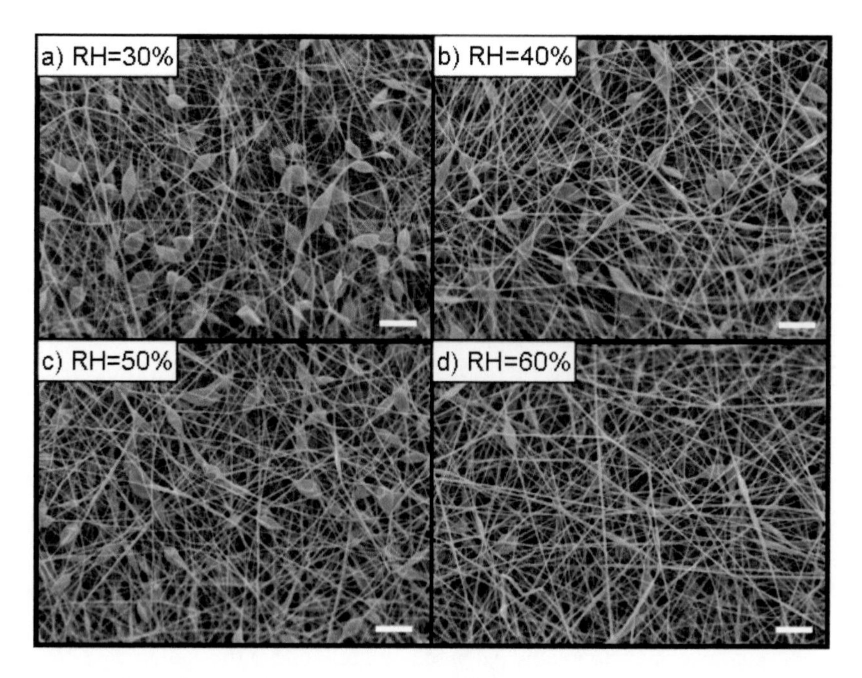

Figure 2. Effect of relative humidity on P(L)LA beaded fibre SEM micrographs of PLLA fibres electrospun at a) RH = 30%, b) RH = 40%, c) RH = 50% and d) RH = 60%, (Electrospinning process conditions: needle diameter = 0.84 mm, applied voltage = 12 kV, needle to collector distance = 15 cm, solution flow-rate = 15 x 10^{-3} ml/min). Scale bar = 10 μm.

Example 2. In order to evaluate the effect of humidity on fibre diameter, a 13% (wt/v) solution of P(L)LA in 65:35 v/v dichloromethane: dimethyl-formamide was used. The polymer concentration was increased from 11% (used in the previous example) to 13% with the aim to obtain fibres without beads.

Indeed, polymer concentration is a parameter that controls solution viscosity and number of chain entanglements, thus regulating both the presence of beads and fibre diameters [35,36]. The temperature was maintained at (20 ± 1 °C) and the relative humidity was changed gradually from 35% to 55%. SEM images of P(L)LA fibres obtained at different humidity levels are reported in Figure 3, together with fibre diameter distribution A broad diameter distribution (860 ± 370 nm) is generated at low humidity (RH = 35%). The distribution becomes narrower with the increase of moisture and shifts towards lower mean values (660 ± 180 nm at RH = 45%). A further increase of humidity broadens the distribution and increases again the fibre diameter (850 ± 260 nm at RH = 55%).

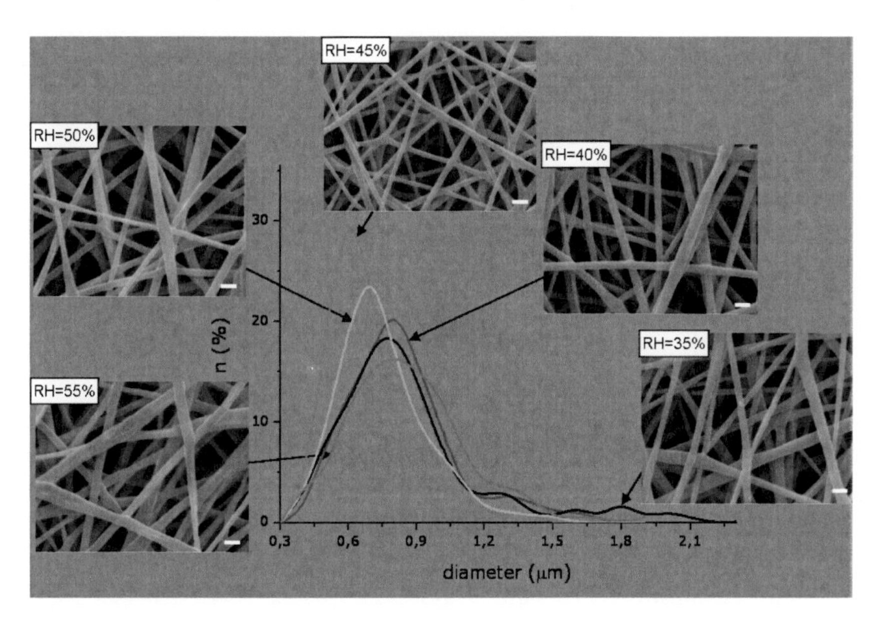

Figure 3. Effect of relative humidity on P(L)LA fibre diameter (solution: 13% (w/v) in 65:35 v/v dichloromethane:dimethylformammide). SEM micrographs and fibre diameter distribution obtained at RH = 35% (black), RH = 40% (blue), RH = 45% (red), RH = 50% (green) and RH = 55% (pink). (Electrospinning process conditions: needle diameter = 0.84 mm, applied voltage = 12 kV, needle to collector distance = 15 cm, solution flow-rate = 15 x 10^{-3} ml/min). Scale bar = 2 μm.

The above reported examples clearly show that control of environmental parameters with a proper instrumental setup is needed in order to achieve electrospinning process reproducibility.

4. Scaffold Wetting

It is well known that electrospun mats are highly hydrophobic, with higher water contact angle values than those of films made of the same materials. Moreover, it has been found that water contact angle decreases with increasing fibre diameter and pore size [63-66].

These results are the consequence of the nano-scale dimension of the mat pores and they are predicted by the Cassie-Baxter theory that relates the different water contact angles of the bulk and of the electrospun mat surface to the air entrapped in the structure [67,68].

Electrospun mats made of the biodegradable polymers used in tissue engineering applications (Table 1) usually float in water or in medium culture (Figure 4.a), since the liquid cannot penetrate the pores which are filled with air. The wettability of electrospun meshes can be reasonably correlated to:

a) the intrinsic hydrophobicity of the material (e.g. PCL is more slowly wetted than PGA),
b) fibre dimension and pore size: electrospun mats made of thick fibres (large mat pores) are more easily wetted than mats made of thin fibres (small mat pores),
c) mat thickness: a thicker mat is more slowly wetted than a thinner one.

Obviously, mat wetting is an essential requirement for scaffold characterization (e.g. mechanical properties in simulated physiological conditions, etc…), in hydrolytic degradation studies, in cell culture, and for fluid exchange through the porous structure.

In order to ensure a fast and complete wetting of the scaffold, air in the pores must be substituted by liquid. To this aim, electrospun scaffolds can be placed in a closed vial containing a liquid (e.g. phosphate buffer solution, culture medium...) and connected with a vacuum system through a needle. While the entrapped air escapes from the mesh, the liquid fills the empty pores and the initially floating mat drops to the vial bottom at the end of the evacuation procedure (Figure 4.b).

Figure 4. Electrospun mat of PLGA (GA content = 50 mol%) a) floating at the surface in water solution and b) dropped to the bottom of the vial after the evacuation procedure.

Scaffold wetting can be also achieved using a liquid that can spontaneously enter the pores and subsequently exchanging it with buffer or culture medium. Ethanol (EtOH), that was successfully used by Mikos. et al. for wetting polylactic acid-based foams [69], can be employed to this aim, since it quickly wets most of the electrospun mats employed in tissue engineering without dissolving them.

5. Scaffold Sterilization

Sterilization of the scaffold is a necessary step prior to any cell culture experiment. A number of techniques are available for biomaterial sterilization. Each of these techniques has peculiar characteristics that may be advantageous or disadvantageous, depending on the polymeric scaffold considered. Thermal treatments may be used for sterilization. Autoclaving is the most commonly employed. The sterilization process consists in gradually increasing water vapour temperature until approximately 120 °C. While this temperature grants for the death of many bacteria it can be detrimental for several biodegradable polymers that usually have melting temperature (T_m) around this value and for polymers susceptible of hydrolytic degradation. For those polymers that can resist above 100 °C, the long exposure to high temperatures may still affect electrospun fibre morphology leading to fibre collapse if polymer T_g is low. A similar sterilization process that involves lower temperatures is chemical vapour sterilization. Among chemicals that can be used, ethylene oxide (EtO) has proven to be safe for biomedical applications. Yet, as this treatment is sometimes performed around 50-60 °C, it is not optimal for polymers with T_g values in this temperature range as shown in Figure 5. for a PLGA (GA content = 50 mol%) electrospun mat. Furthermore, long aeration is required to remove toxic remnants of the chemical vapours used and care should be employed to avoid possible non-desired cross-linking reactions between polymers chains and EtO. Another possibility for sterilization involves irradiation. The easiest method consists in using ultraviolet (UV) light. Typically, the UV lamp built in any common tissue culture hood is able to sterilize porous biomaterials without inducing secondary reactions like cross-linking. Scaffolds are exposed over night (typically 12-16 hours) and wetted with culture medium before cell seeding. Despite the simplicity of this technique, its efficacy is limited since UV radiation has a low penetrating ability and therefore such method is not appropriate for thick mats. UV sterilization efficacy can be improved by reflecting surfaces, but it is still hard to achieve complete sterilization for 3D thick scaffolds. A different irradiation source is γ-rays. This source is commonly used for the sterilization of many medical devices with limited exposure doses (typically between 17 and 25 KGy). Sterilization is complete as γ-rays are able to penetrate into the irradiated materials. Possible concerns arise from potential alterations of the irradiated polymer which can undergo cross-linking reactions , oxidation, or ageing. The simplest and probably the most effective sterilization method for commonly used bioresorbable polymers is ethanol wetting. Electrospun meshes are simply immersed in ethanol for 15-30 minutes and extensively washed before use, typically with phosphate buffered saline solution or cell culture medium. This treatment allows also scaffold wetting (see Paragraph 4). It may be disadvantageous for those polymers whose properties can be negatively affected by ethanol, although there has been no study reporting this for common biomaterials.

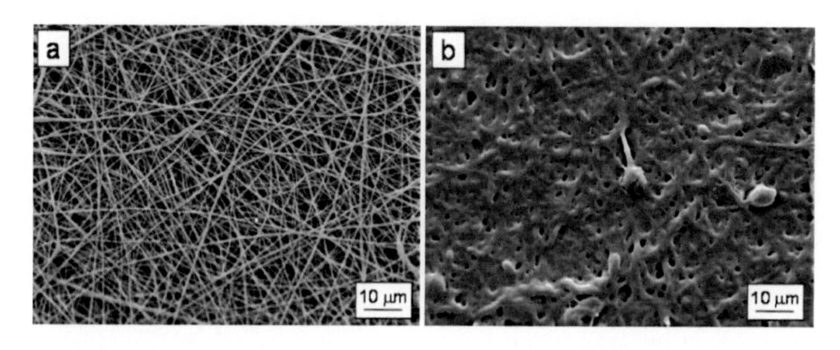

Figure 5. Electrospun mat of PLGA (GA content = 50 mol%) a) as-spun and b) after EtO sterilization at 50 °C.

6. SCAFFOLD SHRINKAGE

Researchers working with electrospun scaffolds are aware that some electrospun mats are not dimensionally stable under cell culture conditions. Indeed, such scaffolds may undergo a macroscopic shrinkage that is accompanied by microscopic changes of fibre morphology: fibres become curly, fibre diameter increases and pore size decreases. This behaviour has been reported for some polymers such as PLGA (GA content = 50 mol%) or P(D,L)LA, whereas other materials such as PCL or P(L)LA do not undergo shrinkage in the same experimental conditions [44,70,71].

In electrospun scaffolds, the macroscopic shrinkage and the change of fibre morphology has been attributed to changes of molecular conformation due to chain relaxation occurring when macromolecules in the amorphous state acquire mobility [44]. Indeed, when the fibre is generated during the electrospinning process, polymer chains are stretched in the fibre axis direction, while the solvent quickly evaporates. If the stretched molecular chains do not have enough time to undergo relaxation before complete solvent evaporation, they will solidify in an elongated conformation. Afterwards, if for any reason macromolecules acquire mobility, they will change their conformation from the stretched oriented one towards the thermodynamically stable random coil one. Accordingly, the macroscopic shrinkage observed will depend on (i) the degree of chain orientation after solvent evaporation and on (ii) the increase of chain mobility in the solid fibre. It is well known that macromolecules in the amorphous state acquire mobility at a temperature close to or higher than their T_g. Polymers usually employed in tissue engineering applications (Table 1) are therefore expected to undergo shrinkage under culture conditions (i.e. culture medium at 37 °C), when their T_g is close to or lower than 37 °C. Moreover, when the same polymers are placed in EtOH, often employed for mat wetting and sterilization (see Paragraphs 4 and 5), the shrinkage can occur even at room temperature since EtOH, that acts as a plasticizer, decreases the polymer T_g below room temperature (see Paragraph 2).

As an example, a square electrospun mat (30 x 30 mm) of P(LA-*co*-TMC) (TMC content = 30 mol%) is placed in EtOH for 1 h at room temperature. The T_g of P(LA-*co*-TMC) electrospun mat is around 34 °C and it decreases down to 0 °C when the mat is kept in EtOH.

Figure 6 shows the shrinkage effect on both random and aligned P(LA-*co*-TMC) fibres together with the percentage of shrinkage calculated as:

$$s = \frac{\ell - \ell}{\ell} \cdot 100$$

Where ℓ is the side length of the mat after EtOH treatment and ℓ is the initial side length (30 mm). After 1 h in EtOH, fibre morphology changes in both types of fibre mats: fibre diameter increases, fibres become more packed and pore dimension decreases (compare Figure 6.a. with Figure 6.c. for random fibres and Figure 6.b. with Figure 6.d. for aligned fibres). As already pointed out, this finding can be attributed to chain relaxation occurring during EtOH treatment. Moreover, the initially aligned fibres loose completely their orientation after EtOH treatment. As regards the degree of shrinkage, in the case of random fibres the two sides of the mat decrease by the same amount ($s_1 = s_2 = -30\%$) due to isotropy of the mesh. Conversely, in the case of the anisotropic mat made of aligned fibres shrinkage of the side parallel to fibre direction (s_p) is larger than shrinkage of the side transversal to fibre direction (s_t) ($s_p = -53\%$ vs $s_t = -30\%$). The above described changes of scaffold dimension and fibre morphology introduce obvious limitations in the use of electrospun materials that undergo strong shrinkage and some authors have even excluded the use of such polymers for tissue engineering applications [44,70,71]. This drawback can be circumvented by a practical trick that allows the macromolecules to relax, while preventing shrinkage. With this aim in mind, the electrospun scaffold is attached to a rigid plastic frame and placed in EtOH. Being the mat bound to the frame, its gross dimensions do not change and fibres tend to maintain their morphology (compare Figures 6.a. with Figure 6.e, for random fibres, and Figures 6.b. with Figure 6.f., for aligned fibres). After this "constrained" pre-treatment in EtOH, the scaffold is removed from the frame and it is immersed again in EtOH to ascertain whether any dimensional changes occur in this second wetting step. Figures 6.g. (random fibres) and 6h (aligned fibres) show that, although some limited shrinkage still occurs, fibre morphology is maintained even if the scaffold is not constrained anymore.

As already pointed out, when P(LA-*co*-TMC) mats are immersed in EtOH, macromolecules undergo transition from the frozen, glassy state to the mobile state. Therefore, results obtained using the above described 'trick' may be interpreted as follows. During the "free" EtOH treatment (i.e. when the scaffold is not 'constrained' by the rigid frame) macromolecules relax from the stretched conformation and undergo spontaneous coiling. A change of fibre morphology and mat dimensions is therefore observed (Figure 6.c. and

Figure 6.d.). These changes do not occur when the mat is immersed in ethanol being attached to a rigid frame (constrained treatment) (Figure 6.e. and Figure 6.f). It should be pointed out that also in this case chain relaxation phenomenon occurs (i.e. macromolecules partially change their conformation from the aligned to the entropically favoured coiled one) but chains also tend to flow one respect to the other, since the constrained fibres have fixed length and they cannot follow the change of molecular conformation. Afterwards, when the scaffold is immersed again in EtOH, without any constraint, only some residual shrinkage is observed (Figure 6.g. and Figure 6.h), that is attributed to a minor fraction of chains still in the stretched oriented conformation. It is reasonable to assume that a longer constrained treatment in EtOH would have allow to completely eliminate any residual shrinkage. Therefore the constrained pre-treatment illustrated above can be an effective way to limit or, after optimization, totally eliminate scaffold shrinkage and fibre morphology changes and may broaden the range of electrospun polymers that can be used for tissue engineering applications.

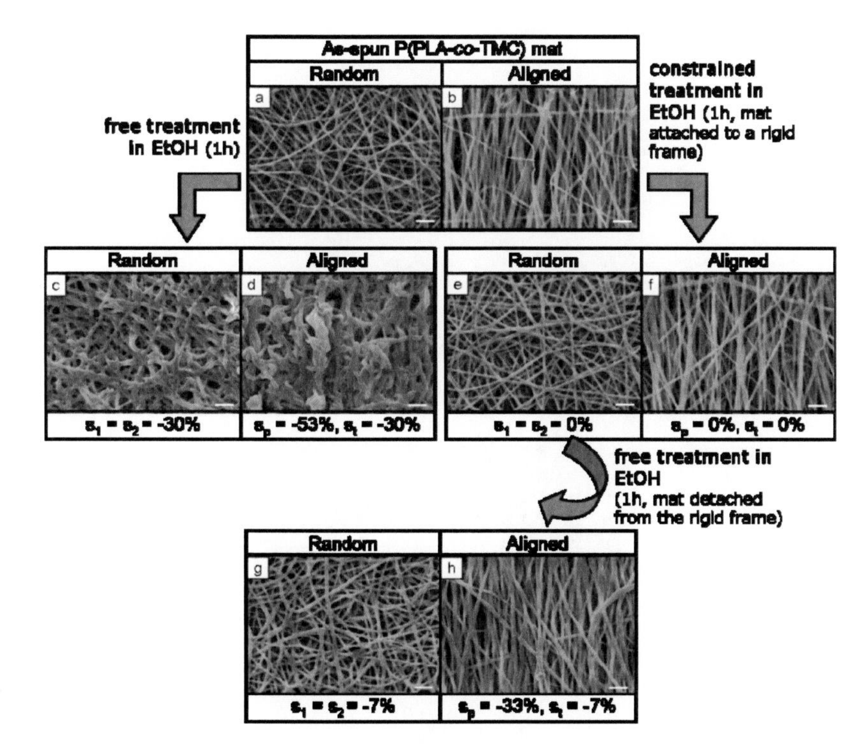

Figure 6. Effect of EtOH treatments on P(LA-*co*-TMC) fibre morphology. Scale bar = 5 μm.

7. Handling of Scaffolds for Cell Culture

As electrospun scaffolds are typically thin mats and consist of extremely light nano or micro fibrillar networks, it is practically difficult to handle them during cell culture experiments. These meshes can easily twist, wind up, and fold. Residual electrostatic charge can also induce sudden displacement of the scaffolds from the place where they were deposited. Using an electrostatic gun helps to partially neutralize the residual net charge on the electrospun scaffolds.

After wetting and sterilization treatments, when electrospun scaffolds are placed in wells containing culture medium, they do not adhere to the well bottom. This often results in loss of efficiency during cell seeding, as cells can attach to the plastic beneath the scaffold, which is normally treated for tissue culture. A rational improvement of seeding efficiency on the scaffold consists in using plastic dishes that are non-treated for tissue culture. Nevertheless, cells can be left in suspension in the medium if electrospun scaffold floating is not avoided. A commercial solution to this problem is offered by Scaffdex, which sells crown rings to immobilise a sample in its place. These inserts are typically custom-made to fit in common culture well-plates with different dimensions. A simpler laboratory scaled solution is to "sink" the scaffolds with a ring of a sterile and inert material preventing it to float in the medium. In addition, if the scaffold is physically attached to that ring (e.g. with non-toxic silicone), a cell leakage-proof well, with the scaffold fixed at the bottom, is obtained (Figure 7). This construct can be inserted in the typical culture well and quantitative cell culture experiments can be carried out. This construct is efficient also in preventing the shrinkage that some electrospun polymers undergo in culture conditions (see Paragraph 6).

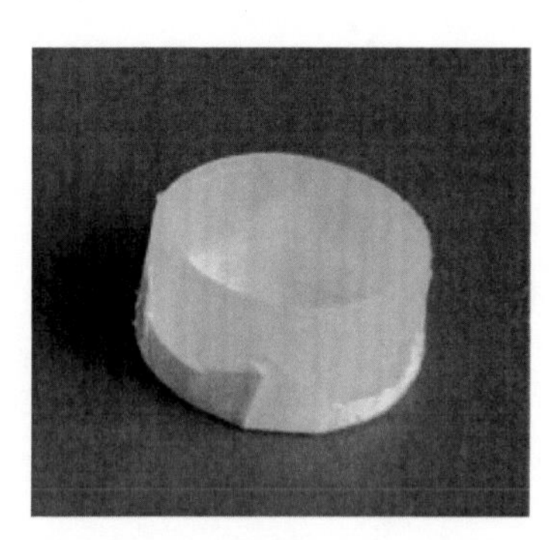

Figure 7. Electrospun scaffold attached to a plastic ring for cell culture experiments.

Conclusion

Electrospun scaffolds hold a tremendous potential for successful applications in tissue engineering as they possess ECM mimicking features in the micro and nano scale ranges that enhance cell-material interactions and tissue regeneration. These cues are either related to fibre diameter or surface topology of the fabricated meshes. Reproducibility of electrospun scaffolds is often a matter of meticulous optimization of different solution, processing, and environmental parameters. When this is achieved, further complications in scaffold handling and use may arise from inappropriate sterilization and post processing treatments, like wetting and thermal/chemical annealing, which can result in fibre morphology variation. In this chapter, we have highlighted some practical hints to better control fabrication reproducibility through an adequate control of environmental processing conditions. In particular, atmosphere humidity is a critical factor to take under control as it can severely affect the reproducibility of fibre morphology. We have also provided some suggestions in order to minimize scaffold shrinkage and to ensure complete scaffold wetting. Finally, scaffold handling during cell culture experiments has been discussed. The problem-solving suggestions contained in this chapter stem from a thorough knowledge of the physical and chemical properties of polymers, a view point which has always to be taken in great consideration when designing scaffolds for regenerative medicine.

References

[1] Langer, R.; Vacanti, J. P. *Science*, 1993, *260*, 920-926.

[2] Vacanti, J. P.; Morse, M. A.; Saltzman, W. M.; Domb, A. J.; Perez-Atayade, A.; Langer, R. *Journal of Pediatric Surgery*, 1988, *23*, 3-9.

[3] Elisseeff, J.; Ferran, A.; Hwang, S.; Varghese, S.; Zhang, Z. *Stem Cells and Development*, 2006, *15*, 295-303.

[4] Lanza, R.; Langer, R.; Vacanti, *J. Principles of Tissue Engineering*; Elsevier Academic Press, 2007.

[5] Ikada, Y. *Tissue Engineering: Fundamentals and Applications*; Elsevier Academic Press, 2006.

[6] Schoof, H.; Apel, J.; Heschel, I.; Rau, G. *Journal of Biomedical Material Research Part B: Applied Biomaterials*, 2001, *58*, 352-357.

[7] Hollister, S. J. *Nature Materials*, 2005, *4*, 518-524.

[8] Hutmacher, D. W. *Journal of Biomaterials Science, Polymer Edition*, 2001, *12*, 107-124.

[9] Li, D.; Xia, Y. *Advanced Materials*, 2004, *16*, 1151-1170.

[10] Elisseeff, J.; Puleo, C.; Yang, F.; Sharma, B. *Orthodontics and Craniofacial Research*, 2005, *8*, 150-161.

[11] Murphy, M. B.; Mikos, A. G. In *Principles of Tissue Engineering*, 3rd ed.; Lanza, R., Langer, R., Vacanti, J. Eds.; Elsevier Academic Press, 2007.

[12] Bognitzki, M.; Czago, W.; Frese, T.; Schaper, A.; Hellwing, M.; Steinhart, M.; Greiner, A.; Wendorff, J. H. *Advanced Materials*, 2001, *13*, 70-72.

[13] Buttafoco, L.; Kolkman, N. G.; Engbers-Buijtenhuijs, P.; Poot, A. A.; Dijkstra, P. J.; Vermes, I.; Feijen, J. *Biomaterials*, 2006, *27*, 724-734.

[14] Liao, S.; Li, B.; Ma, Z.; Wei, H.; Chan, C.; Ramakrishna, S. *Biomedical Materials*, 2006, *1*, R45-R53.

[15] Teo, W.-E.; Ramakrishna, S. *Nanotechnology*, 2006, *17*, R89-R106.

[16] Rutledge, G.; Li, Y.; Fridrikh, S.; Warner, S.; Kalayci, V.; Patra, P. *National Textile Center Annual Report*, 2001, 1-10.

[17] Deitzel, J. M.; Kleinmeyer, J. D.; Harris, D.; Beck Tan, N. C. *Polymer*, 2001, *42*, 261-272.

[18] Ko, F. *NATO ASI Proceedings*, 2003, 1-18.

[19] Sun, Z.; Zussman, E.; Yarin, A.; Wendorff, J.; Greiner, A. *Advanced Materials*, 2003, *15*, 1929-1932.

[20] Kidoaki, S.; Kwon, I. K.; Matsuda, T. *Biomaterials*, 2005, *26*, 37-46.

[21] Moroni, L.; Licht, R.; de Boer, J.; de Wijn, J. R.; van Blitterswijk, C. A. *Biomaterials*, 2006, *27*, 4911-4922.

[22] Badami, A. S.; Kreke, M. R.; Thompson, M. S.; Riffle, J. S.; Goldstein, A. S. *Biomaterials*, 2006, *27*, 596-606.

[23] Li, W. J.; Jiang, Y. J.; Tuan, R. S. *Tissue Engineering*, 2006, *15*, 1775-1785.

[24] Jayasinghe, S. N.; Qureshi, A. N.; Eagles, P. A. *Small*, 2006, *2*, 216-219.

[25] Cui, W.; Li, X.; Yu, G.; Zhou, S.; Weng, J. *Biomacromolecules*, 2006, *7*, 1623-1629.

[26] Jiang, H.; Hu, Y.; Li, Y.; Zhao, P.; Zhu, K.; Chen, W. *Journal of Controlled Release*, 2005, *108*, 237-243.

[27] Li, C.; Vepari, C.; Jin, H. J.; Kim, H. J.; Kaplan, D. L. *Biomaterials*, 2006, *27*, 3115-3124.

[28] Stankus, J. J.; Guan, J.; Fujimoto, K.; Wagner, W. R. *Biomaterials*, 2006, *27*, 735-744.

[29] Xu, C.; Inai, R.; Kotaki, M.; Ramakrishna, S. *Tissue Engineering*, 2004, *10*, 1160-1168.
[30] Min, B. M.; Lee, G.; Kim, S. H.; Nam, Y. S.; Lee, T. S.; Park, W. H. *Biomaterials*, 2004, *25*, 1289-1297.
[31] Sahoo, S.; Ouyang, H.; Goh, J. C.; Tay, T. E.; Toh, S. L. *Tissue Engineering*, 2006, *12*, 91-99.
[32] Li, W. J.; Tuli, R.; Okafor, C.; Derfoul, A.; Danielson, K. G.; Hall, D. J.; Tuan, R. S. *Biomaterials*, 2005, *26*, 599-609.
[33] Tuzlakoglu, K.; Bolgen, N.; Salgado, A. J.; Gomes, M. E.; Piskin, E.; Reis, R. L. *Journal of Materials Science: Materials in Medicine*, 2005, *16*, 1099-1104.
[34] Moroni, L.; Schotel, R.; hamann, D.; de Wijn, J. R.; van Blitterswijk, C. A. *Advanced Functional Materials*, 2008, *18*, 53-60.
[35] Gupta, P.; Elkins, C.; Long, T. E.; Wilkes, G. L. *Polymer*, 2005, *46*, 4799-4810.
[36] Shenoy, S. L.; Bates, W. D.; Frisch, H. L.; Wnek, G. E. *Polymer*, 2005, *46*, 3372-3384.
[37] Shum, A. W. T.; Mak, A. F. T. *Polymer Degradation and Stability*, 2003, *81*, 141-149.
[38] You, Y.; Min, B.-M.; Lee, S. J.; Lee, T. S.; Park, W. H. *Journal of Applied Polymer Science*, 2005, *95*, 193-200.
[39] You, Y.; Lee, S. W.; Lee, S. J.; Park, W. H. *Materials Letters*, 2006, *60*, 1331-1333.
[40] Inai, R.; Kotaki, M.; Ramakrishna, S. *Nanotechnology*, 2005, *16*, 208-213.
[41] Zhong, X.; Kim, K.; Fang, D.; Ran, S.; Hsiao, B. S.; Chu, B. *Polymer*, 2002, *43*, 4403-4412.
[42] Ren, J.; Liu, W.; Zhu, J.; Gu, S. *Journal of Applied Polymer Science*, 2008, *109*, 3390-3397.
[43] Tan, E. P. S.; Lim, C. T. *Nanotechnology*, 2006, *17*, 2649-2654.
[44] Zong, X.; Ran, S.; Kim, K.-S.; Fang, D.; Hsiao, B. S.; Chu, B. *Biomacromolecules*, 2003, *4*, 416-423.
[45] Zhong, X.; Ran, S.; Fang, D.; Hsiao, B. S.; Chu, B. *Polymer*, 2003, *44*, 4959-4967.
[46] Zeng, J.; Chen, X.; Liang, Q.; Xu, X.; Jing, X. *Macromolecular Bioscience*, 2004, *4*, 1118-1125.
[47] Sombatmankhong, K.; Suwantong, O.; Waleetorncheepsawat, S.; Supaphol, P. *Journal of Polymer Science: Part B: Polymer Physics*, 2006, *44*, 2923-2933.

[48] Kim, G.-M.; Michler, G. H.; Henning, S.; Radush, H.-J.; Wutzler, A. *Journal of Applied Polymer Science*, 2007, *103*, 1860-1867.
[49] Gilding, D. K.; Reed, A. M. *Polymer*, 1979, *20*, 1459-1464.
[50] Engelberg, I.; Kohn, J. *Biomaterials*, 1991, *12*, 292-304.
[51] Lou, X.; Detrembleur, C.; Jerome, R. *Macromolecular Rapid Communications*, 2003, *24*, 161-172.
[52] Brandrup, J.; Immergut, E. H.; Grulke, E. A. *Polymer Handbook*; Wiley-Interscience Publication, 1999.
[53] Deitzel, J. M.; Kleinmeyer, J. D.; Hirvonen, J. K.; Beck Tan, N. C. *Polymer*, 2001, *42*, 8163-8170.
[54] Andrady, A. L. *Science and Technology of Polymer Nanofibers*; John Wiley and Sons, Inc., 2008.
[55] Ramakrishna, S.; Fujihara, K.; Teo, W.-E.; Lim, T.; Ma, Z. *An Introduction to Electrospinning and Nanofibers*; World Scientific Publishing, 2005.
[56] de Vrieze, S.; Van Camp, T.; Hagstrom, B.; Westbroek, P.; de Clerck, K. *Journal of Materials Science*, 2008,
[57] Megelski, S.; Stephens, J. S.; Chase, D. B.; Rabolt, J. F. *Macromolecules*, 2002, *35*, 8456-8466.
[58] Casper, C. L.; Stephens, J. S.; Tassi, N. G.; Chase, D. B.; Rabolt, J. F. *Macromolecules*, 2004, *37*, 573-578.
[59] Kongkhlang, T.; Kotaki, M.; Kousaka, Y.; Umemura, T.; Nayaka, D.; Chirachanchai, S. *Macromolecules*, 2008, *41*, 4746-4752.
[60] Tripatanasuwan, S.; Zhong, Z.; Reneker, D. H. *Polymer*, 2007, *48*, 5742-5746.
[61] Medeiros, E. S.; Mattoso, L. H. C.; Offeman, R. D.; Wood, D. F.; Orts, W. J. *Canadian Journal of Chemistry*, 2008, *86*, 590-599.
[62] Kadomae, Y.; Amagasa, M.; Sugimoto, M.; Taniguchi, T.; Koyama, K. *International Polymer Processing*, 2008, *23*, 377-384.
[63] Cui, W.; Li, X.; Zhou, S.; Weng, J. *Polymer Degradation and Stability*, 2008, *93*, 731-738.
[64] Huang, F.; Wei, Q.; Cai, Y.; Wu, N. *International Journal of Polymer Analysis and Characterization*, 2008, *13*, 292-301.
[65] Yoon, Y.; Moon, H. S.; Lyoo, W. S.; Lee, T. S.; Park, W. H. *Carbohydrate Polymers*, 2009, *75*, 246-250.
[66] Wu, W.; Zhu, Q.; Qing, F.; Han, C. C. *Langmuir*, 2009, *25*, 17-20.
[67] Cassie, B. D.; Baxter, S. *Transactions of the Faraday Society*, 1944, *40*, 546-551.

[68] Acatay, K.; Simsek, E.; Yang, C. O.; Menceloglu, Y. Z. *Angewandte Chemie International Edition*, 2004, *43*, 5210-5213.

[69] Mikos, A. G.; Lyman, M. D.; Freed, L. E.; Langer, R. *Biomaterials*, 1994, *15*, 55-58.

[70] Li, W. J.; Cooper, J.; Mauck, R. L.; Tuan, R. S. *Acta Biomaterialia*, 2006, *2*, 377-385.

[71] Hong, Y.; Fujimoto, K.; Hashizume, R.; Guan, J.; Stankus, J. J.; Tobita, K.; Wagner, W. R. *Biomacromolecules*, 2008, *9*, 1200-1207.

In: Electrospinning Process and Nanofiber… ISBN 978-1-61209-330-7
Editors: A.K. Haghi and G.E. Zaikov

Chapter 10

Preparation and Structures of Electrospun PAN Nanofibers

*A.K. Haghi**
Textile Engineering Department, Faculty of Engineering,
University of Guilan, Rasht, Iran

Abstract

A simple and non-conventional electrospinning technique was employed for producing highly oriented Polyacrylonitrile (PAN) nanofibers. The PAN nanofibers were electrospun from 14 wt% solution of PAN in dimethylformamid (DMF) at 11 kv on a rotating drum with various linear speeds from 22.5 m/min to 67.7 m/min. The influence of take up velocity was investigated on the degree of alignment, internal structure and mechanical properties of collected PAN nanofibers. Using an image processing technique, the best degree of alignment was obtained for those nanofibers collected at a take up velocity of 59.5 m/min. Moreover, Raman spectroscopy was used for measuring molecular orientation of PAN nanofibers. Similarly, a maximum chain orientation parameter of 0.25 was determined for nanofibers collected at a take up velocity of 59.5 m/min.

Keywords: Polyacrylonitrile, Nanofiber, Orientation, Electrospinning.

---

* E-mail: Haghi@guilan.ac.ir, Tel: +98-131-6690270, Fax: +98-131-6690271

INTRODUCTION

With potential applications ranging from protective clothing, tissue engineering and filtration technology to reinforcement of composite nanomaterials, nanofibers offer a remarkable opportunity toward development of multifunctional nanostructural systems (Huang et al., 2003; Ramakrishna et al., 2005; Fennessey et al., 2004; Pan, 2006).

The emergence of various applications is inspired by outstanding properties of nanofibers such as huge surface area per mass ratio (Huang et al., 2003) and high porosity along with small pore size (Ramakrishna et al., 2005). Moreover, for diverse applications, highly oriented and flexible nanofiber with superior mechanical properties is extremely demanded.

The electrospinning process is a sophisticated technique for producing nanofibers based on applying a high voltage DC electric potential between the end of a capillary tube and a collector. When the applied electric field overcomes the surface tension of the droplet, a charged jet of polymer solution is ejected and nanofibers are collected on the target (Fennessey et al., 2004). Recent studies have shown that aligned nanofibers have better molecular orientation and, as a consequence, improved mechanical properties than randomly oriented nanofibers (Fennessey et al., 2004; Zussman et al., 2005; Gu et al., 2005).

Additionally, the aligned nanofibers are better suited for preparing carbon nanofibers from electrospun PAN nanofiber precursors (Jalili et al., 2006). In another attempt (Fennessey et al., 2004) tows of unidirectional and molecularly oriented PAN nanofibers were prepared using a high speed, rotating take up wheel.

A maximum orientation factor of 0.23 was determined for nanofibers collected between 8.1 m/s and 9.8 m/s. The aligned tows were twisted into yarns, and the mechanical properties of the yarns were determined as a function of twist angle. Their produced yarn with twist angle of 11° had an initial modulus and ultimate strength of about 5.8 GPa and 163 MPa, respectively (Fennessey et al., 2004). Zussman et al. (2005) have demonstrated the use of a wheel-like bobbin as the collector to position and align individual PAN nanofibers into parallel arrays with an orientation factor of 0.34 at a collection speed of 5 m/s. In another study, aligned PAN nanofibers collected across the gap between the two grounded strips of aluminum foil showed an obvious improvement of mechanical properties in the modulus of the resultant carbon nanofibers, as a consequence of an increase in the orientation factor from 0 to 0.127 (Gu et al., 2005).

The current study focuses on preparation of aligned and molecularly oriented PAN nanofibers using two needle pumps in a highly productive approach.

Various take up velocities were examined to obtain the highest possible collection speed capable of producing aligned nanofiber. Non-destructive techniques like angular power spectra (APS) and Raman spectroscopy were utilized for characterization of alignment and molecular orientation in nanofibers.

EXPERIMENTAL

Material and Reagents

Industrial polyacrylonitrile (PAN) with average molecular weight of 100000 g/mol and highly pure dimethylformamide (DMF, 99%) were supplied respectively by Iran Polyacryle and Merck companies. All PAN/DMF solutions were prepared using constant power magnetic stirrer at room temperature.

Electrospinning Setup

The electrospinning apparatus consists of a high voltage power supply; two syringe pumps, two stainless steel needles (0.7 mm OD) and a rotating collector equipped to an inverter for controlling linear speed (Figure 1).

In this setup, unlike the conventional technique, two needles were installed in opposite directions and polymer solutions were pumped to needles by two syringe infusion pumps with same injection rate.

The flow rate of solutions to the needle tip is maintained constant so that a pendant drop remains during electrospinning. The horizontal distance between the needles and the collector was 20-25 cm. When high voltages (9-11 KV) were applied to the needles with opposite voltage, jets were ejected simultaneously.

Then the jets with opposite charges attracted each other, stuck together and a cluster of fibers is formed. For collecting aligned nanofibers, the cluster of fibers formed between the two needles was towed manually to a rotating drum.

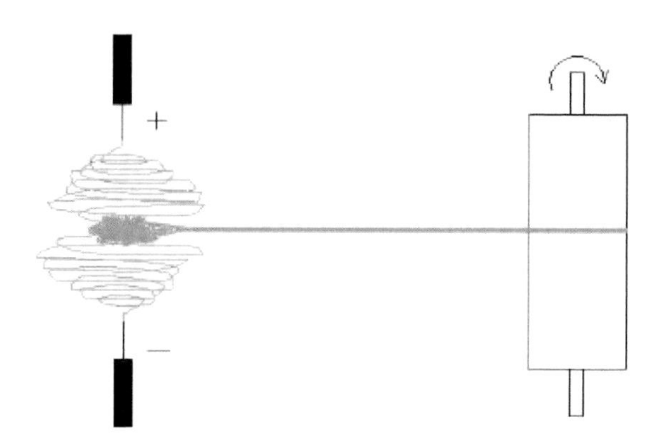

Figure 1. Schematic of electrospinning setup for collection of continuous aligned nanofiber.

Characterization Techniques

Various characterization techniques were applied to 10 nanofiber samples produced in similar conditions. The geometry of electrospun nanofibers was investigated using a Philips XL-30 scanning electron microscope (SEM). Nanofiber samples were mounted onto SEM plates; sputter coated with gold and inserted in a SEM vacuum vessel for image capturing. Also, the Motic optical microscope was employed to acquire images for alignment analysis. Fiber alignment was characterized using image-processing technique by Fourier Power Spectrum (FPS) as a function of take up speed. All images used in the process were obtained at a resolution of 640×480 pixels and 1000 times magnifications. The number of captured images was 30 at each of the take up speeds. The variation of molecular orientation with respect to take up speed was characterized by Polarized Raman spectroscopy. Raman spectra were obtained with a Thermo Nicolt Raman Spectrometer model Almega Dispersive 5555. The spectra were collected in the backscattering mode, using the 532 nm line of a Helium/Neon laser. The nominal power of the laser was 30 mW. The nanofiber axis is oriented at an angle of $\theta°$ with respect to the machine direction. The angle between the polarization plane and the nanofiber axis is ψ. The orientation studies were performed when fibers were at $\psi = 0°$ and $\psi = 90°$ to the plane of polarization of the incident laser. At each angle, the enhanced spectra at VV (polarized laser and the analyzer are parallel) and VH (the polarized laser and the analyzer are perpendicular) configurations were collected.

RESULTS AND DISCUSSION

Well-aligned PAN nanofibers with very long lengths were electrospun using a simple and unconventional method. In this technique, the electric field only exists between two syringe needles (not between needle ad collector) and, thus, the movement of nanofibers in the drawing area (distance between the needles and the collector) is due to mechanical force caused by the rotating drum and electric field and therefore has no impact on the drawing mechanism.

Various operational and material parameters were adjusted in order to obtain optimum and productive conditions. PAN nanofibers prepared at 14 and 15 wt% have lower rupture compared to 13 wt% polymer solution, presumably due to a higher level of chain entanglement. The conditions pertinent to minimum number of rupture for nanofibers prepared at different concentrations are shown in Table 1.

Figure 2 shows SEM images of PAN nanofibers electrospun in optimum conditions. The generated nanofibers have uniform structure without any bead formation. The average diameter of nanofibers increases with increasing polymer concentration because of lower extension of the jet at higher concentration and viscosity. Based on SEM images, the average diameter of nanofibers produced from 13 wt% polymer solutions is about 323 ± 31 nm, which is about 100 nm less than the diameter of those fibers collected from 15 wt% polymer solution (Ziabari et al., 2009). This is probably due to the greater stretching resistance of the solution with increasing solid content (Ramakrishna et al., 2005). However the 14 wt% concentration was chosen for producing of PAN nanofibers due to a lower number of rupture and higher diameter uniformity.

Table 1. Various conditions used for nanofiber preparation

Solution concentration (wt%)	Applied voltage (kv)	Feed rate (mL/h)	Distance between two needles(cm)	Distance between needles and collector (cm)	Total number of rupture at (15 minutes) [1]
13	10.5	0.293	13	20	12
14	11	0.293	13	20	6
15	11	0.293	15	20	6

1. It was measured visually by counting the number of rupture of fiber flow between electrospinning needles and rotating collector.

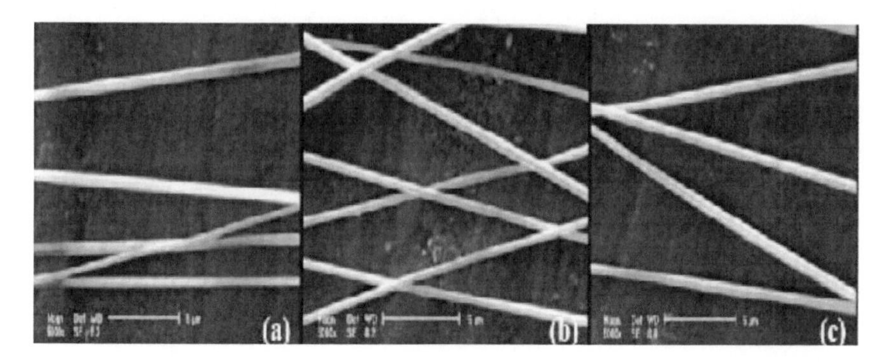

Figure 2. SEM images of PAN nanofibers (a) 13 wt% (b) 14 wt% (c) 15 wt% polymer solution.

Table 2. The degree of alignment at different take-up speeds

Take up speed	22.5	31.6	40.6	49.6	59.5	67.7
Degree of alignment	24.6±3.9	34.4±5.2	32.7±7.65	29.5±5.97	37.5±5.2	29.4±7.1

Alignment analysis of nanofibers collected at different take up speeds from 22.5 m/min to 67.7 m/min was carried out using the angular power spectrum (APS) based on micrographs captured by an optical microscope. Acquired image was processed using MATLAB software to produce angular power spectrum as the square of the Fourier power spectrum. APS is a plot of how much the orientation varies from point to point on the web of nanofibers versus the angular frequency (the x-axis variable). The sharpness of the peak at an angle of 90° on the APS shows the degree of alignment of nanofibers in the vertical direction and the area of that peak shows the density of aligned nanofibers (Jalili et al., 2006). Furthermore, the plot of the normalized APS (ratio of intensity of the APS to the corresponding mean intensity of the Fourier power spectrum) versus angle was used for calculating the degree of alignment (Figure 3). Disorientation of nanofibers generates a broad spectrums around 90°, which reflects directly a low degree of alignment. The ratio of peak area at $90^{\circ}\pm3^{\circ}$ to total area of the APS plot was utilized to calculate the density of aligned nanofibers. Table 2 shows the influence of the linear take up speed of the rotating collector on the degree of alignment of the nanofibers.

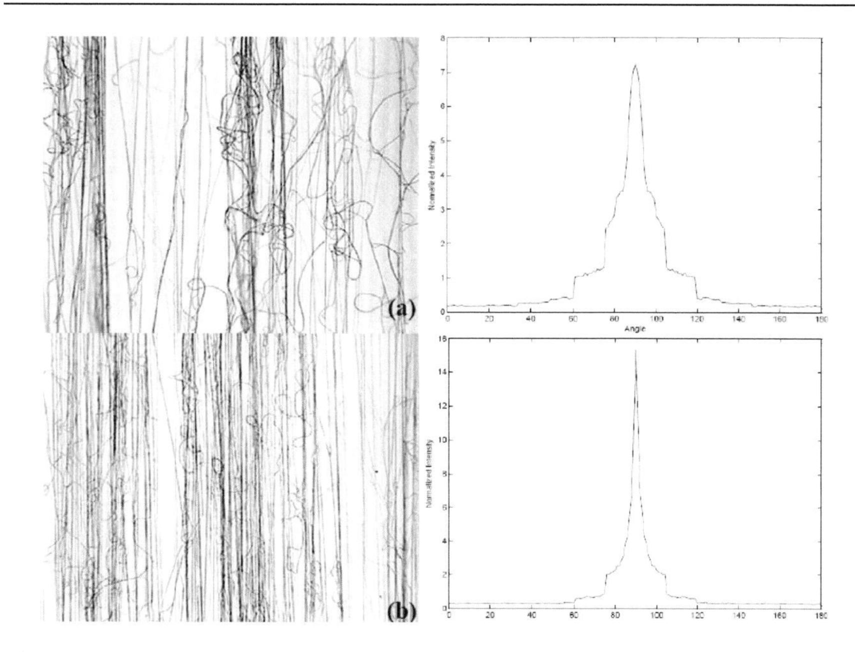

Figure 3. Optical micrograph of electrospun PAN nanofibers with the corresponding normalized APS at take up speeds of (a) 22.5 m/min (b) 59.5 m/min.

Table 3. The intensity ratios of the enhanced peak at 1394 cm^{-1} using VV and VH configuration

Sample	Rotating drum speed (m/min)	IVV0	IVV 90	IVH0	IVV90/ IVH0	IVV0/ IVH0	IVV0/ IVV90
A	22.5	4400	1900	1600	1.19	2.75	2.31
B	49.6	3600	1500	2400	0.63	1.50	2.40
C	59.5	4000	1500	4277	0.35	0.93	2.67
D	67.7	3800	2200	6350	0.35	0.60	1.73

The higher the take up speed, the more aligned is the nanofiber collected. As the rotation speed increases, the effective draw (difference between the surface velocity of the drum and the final velocity of the fiber) is increased, resulting in better alignment of the collected fiber and less deviation between the fiber and rotation direction.

However, the positive impact of linear take up velocity on degree of alignment passes through a maximum of 37.5% at 59.5 m/min and the trend of the data reverses after this collection speed. The decrease of the degree of alignment after a specific take up speed presumably is due to insufficient time for arrangement of the molecular chains in the drawing mechanism.

The drawing mechanism inherently uncoils the molecular chain to reach higher orientation, but after an optimum take up speed the breaking force acting on the molecular chain starts to build up.

At the early stage, the decrease of nanofiber alignment can be observed due to partial segregation of the polymer chains and reforming of the coiled structure. Meanwhile, if the take up speed increase is continued, complete fracture occurs in the nanofiber structure.

The molecular orientation of the nanofibers was also characterized using Polarized Raman Spectroscopy. Figure 4 shows the Raman spectra of nanofiber samples collected at different take up speeds under the VV configuration at $\psi = 0^{o}$ (parallel) and 90^{o}(Vertical) versus the polarization plane.

Four samples with symbols A, B, C and D were prepared using conditions given in Table 1 for 14 wt % polymer solution and various rotating drum speeds. Raman spectra of PAN nanofibers usually arise in the region of 500-1500 cm^{-1} which is commonly observed as the Raman fingerprint of PAN microfiber (Huang and Koenig 1971) .

Compared to the peak enhancement at 600 cm^{-1} with constant intensity at different polarization angles, the intensity at 1394 cm^{-1} monotonically decreases with increasing ψ. The intensity dependence of the peak enhancement at 1394 cm^{-1} on the angle between fiber and polarization plane can be considered as a powerful tool for determination of nanofiber orientation. Other peaks that enhanced between 1394 cm^{-1} and 600cm^{-1} did not decrease significantly, except the peak enhanced at 1190 cm^{-1}.

Similar to the VV configuration, the peak intensity observed for different samples in the VH configuration at ψ=0 characterized the degree of molecular orientation. According to the intensity ratios shown in Table 3, Raman spectra show a much stronger orientation effect in sample (C) compared to other spun nanofibers.

It is clear that the higher take up speed is mostly responsible for the orientation of the molecular chains in the fiber direction. Therefore, as expected, a low orientation dependence of the Raman modes was observed for the lower take up speed.

The orientation parameters of $\langle P2 (\cos\theta)\rangle$ and $\langle P4 (\cos\theta)\rangle$, which are, respectively, the average values of the Legendre polynomials P2 (cosθ) and P4 (cosθ) for the bulk product, can be calculated based on the Raman intensity ratios in VV and VH modes given in Table 3. (Liu and Kumar 2003).

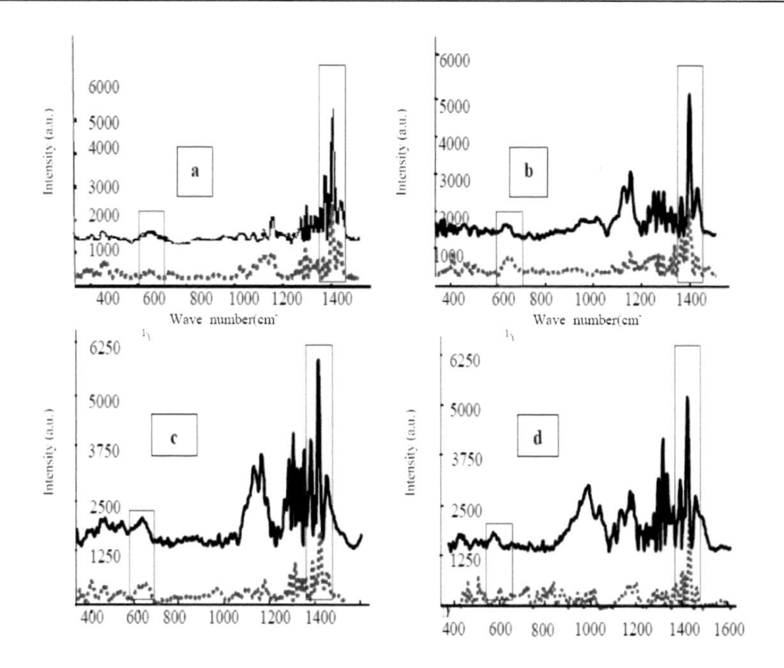

Figure 4. The Raman spectra under the VV mode: (a) 22.5 m/min(b) 49.6 m/min(c) 59.5 m/min(d) 67.7m/min. The angle between the fiber axis and the polarization plane for filled curve and dashed curve are 0 ° and 90 °, respectively.

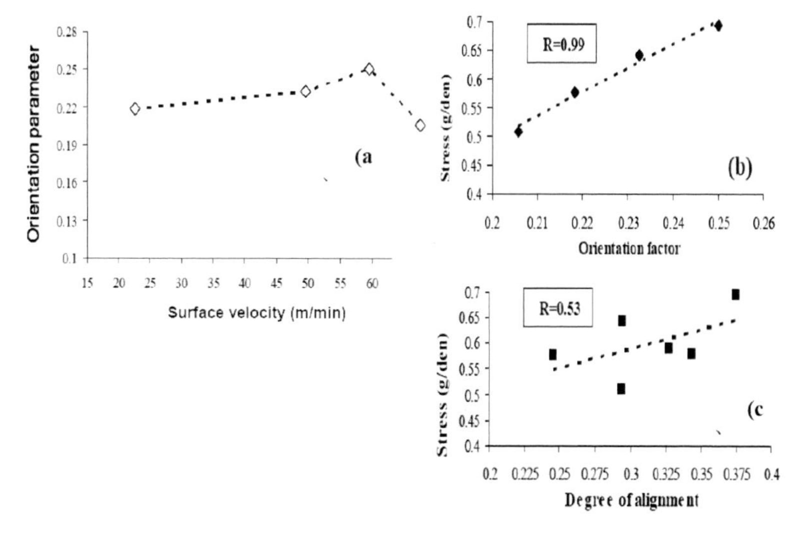

Figure 5. (a)Orientation parameter versus take up speed determined by Raman spectroscopy; (b) Stress versus molecular orientation parameter; (c) Stress versus degree of alignment.

More specifically, <P2(cosθ)> is known as the Herman orientation factor (f), which varies between values of 1 and 0, corresponding, respectively, to a nanofiber fully oriented in the take up direction and a fully non-oriented nanofiber distribution.

The Herman orientation factor for different samples varies between 0.20 and 0.18 when the take up speed increases from 22.5 m/min to 67.7 m/min (Figure 5.a). It can be clearly seen that the degree of orientation reaches a maximum point at 59.5 m/min and then decreases at higher take-up speed. This trend is in agreement with data acquired by optical microscopy through the normalized APS, as shown in previous work (Fennessey et al., 2004).

The orientation parameter decreases for greater take up speeds probably due to the decrease of the time necessary for drawing nanofibers; as a result, there is a lack of sufficient opportunity for arranging the molecular chains in the tension direction.

Comparing the higher molecular orientation parameter (up to 0.25 at a take up speed of 59.5 m/min ~ 1m/s) achieved in the current study with that of other methods shows that this technique had a more effective influence on the molecular orientation of nanofibers than other methods (Fennessey et al., 2004; Zussman et al., 2005).

In other cases in which two grounded strips were used as a collector, the obtained value of the chain orientation parameter (0.127) was less than in our study(Zussman et al., 2005).

On the other hand, using the rotating drum as a collector led to the measured orientation parameter of 0.23 at higher take up speed (9.84 m/s) (Fennessey et al., 2004).

Therefore it can be speculated that, considering the more effective applied stretch in this system, the presented above method will help to reach suitable molecular orientations at lower take up speeds. Figures 5.b.c. represent, respectively, the plots of stress versus molecular orientation parameter and degree of alignment.

The higher the degree of alignment, the greater the number of fibers oriented in the tension direction. The higher correlation factor of 0.99 (stress-orientation factor curve) compared to 0.53 (stress- degree of alignment curve) suggests that the molecular orientation is more responsible for good mechanical properties of the resultant nanofibers than degree of alignment. The former factor is at the molecular level and controls the degree of crystallinity and, therefore, the macroscopic properties.

Conclusion

A simple and non-conventional electrospinning technique using two syringe pumps was employed for producing highly oriented polyacrylonitrile (PAN) multifilament nanofibers. The process was carried out using two needles in opposite positions and a rotating collector perpendicular to the needle axis. The current procedure was optimized for increasing of orientation and productivity of nanofibers with diameters in the nanoscale range. PAN nanofibers were electrospun from 14 wt% solutions of PAN in dimethylformamide (DMF) at 11 kv on a rotating drum with various linear speeds from 22.5 m/min to 67.7 m/min. The influence of take up velocity on the degree of alignment, internal structure and mechanical properties of collected PAN nanofibers was also investigated. Various characterization techniques were employed to find out the influence of operational parameters on the degree of orientation. Based on micrographs captured by optical microscopy, the angular power spectrum (APS) was generated based on an image processing technique. The best degree of alignment was obtained for those nanofibers collected at a take up velocity of 59.5 m/min. Moreover, polarized Raman spectroscopy under VV configuration at $\psi = 0^{\circ}$ (parallel) and 90° (Vertical) versus polarization plane and also in VH configuration at $\psi=0$ was used as a standard technique for measuring molecular orientation of PAN nanofibers. Similarly, a maximum chain orientation parameter of 0.25 was determined for nanofibers collected at a take up velocity of 59.5 m/min.

References

Fennessey, S.F. and Farris, R.J., Fabrication of aligned and molecularly oriented electrospun polyacrylonitrile nanofibers and the mechanical behavior of their twisted yarns, *Polymer*, 45, 4217(2004).

Gu, S.Y., Ren, J. and Wu, Q.L., Preparation and structures of electrospun PAN nanofibers as a precursor of carbon nanofibers, *Synthetic Metals,* 155,157(2005).

Huang, Y.S. and Koenig, J.L.,Raman spectra of polyacrylonitrile, *Applied Spectroscopy*, 25, 620-622(1971).

Huang, Z.M., Zhang, Y.Z., Kotaki, M. and Ramakrishna, S., A review on polymer nanofibers by electrospinning and their applications in nanocomposites, *Composites Science and Technology,* 63, 2223 (2003).

Jalili, R., Morshed, M. and Hosseini Ravandi, S.A., Fundamental Parameters Affecting Electrospinning of PAN Nanofibers as Uniaxially Aligned Fibers, *Journal of Applied Polymer Science*, 101, 4350(2006).

Liu, T. and Kumar, S., Quantitative characterization of SWNT orientation by polarized Raman spectroscopy, *Chem. Phys. Lett,* 378, 257(2003).

Pan, H., Li, L., Hu, L. and Cui, X., Continuous aligned polymer fibers produced by a modified electrospinning method, *Polymer*, 47, 4901(2006).

Ramakrishna, S., Fujihara, K., Teo, W.E., Lim, T.C. and Ma, Z., An introduction to electrospinning and nanofibers, World scientific, Singapore (2005).

Ziabari, M., Mottaghitalab, V. and Haghi, A. K. Application of direct tracking method for measuring electrospun nanofiber diameter. *Braz. J. Chem.* Eng., 26, 1, 53(2009).

Zussman, E., Chen, X., Ding, W., Calabri, L., Dikin, D.A., Quintana, J. P. and Ruoff, R.S., Mechanical and structural characterization of electrospun PAN-derived carbon nanofibers, *Carbon*, 43, 2175(2005).

INDEX

D

E

F

N

O

P

T

U

V

W

X

Y